brown pelican

LOUISIANA TRUE books tell the stories of the state's iconic places, traditions, foods, and objects. Each book centers on one element of Louisiana's culture, unpacking the myths, misconceptions, and historical realities behind everything that makes our state unique, from above-ground cemeteries to zydeco.

---

brown pelican

RIEN FERTEL

Louisiana State University Press

Baton Rouge

Published by Louisiana State University Press
lsupress.org

LSU Press Paperback Original
Manufactured in the United States of America
First printing

Designer: Barbara Neely Bourgoyne
Typeface: Source Sans Variable
Printer and binder: Integrated Books International (IBI)

Cover illustration: iStock.com/Gail Salter (modified)

Library of Congress Cataloging-in-Publication Data
Names: Fertel, Rien, 1980– author.
Title: Brown pelican / Rien Fertel.
Other titles: Louisiana true.
Description: Baton Rouge : Louisiana State University Press, [2022] |
    Series: Louisiana true | Includes bibliographical references.
Identifiers: LCCN 2022007943 (print) | LCCN 2022007944 (ebook) | ISBN
    978-0-8071-7846-1 (paperback) | ISBN 978-0-8071-7880-5 (pdf) | ISBN
    978-0-8071-7879-9 (epub)
Subjects: LCSH: Brown pelican—Louisiana. | Brown pelican—
    Conservation—Louisiana.
Classification: LCC QL696.P47 F47 2022 (print) | LCC QL696.P47 (ebook) | DDC
    598.4/3—dc23/eng/20220223
LC record available at https://lccn.loc.gov/2022007943
LC ebook record available at https://lccn.loc.gov/2022007944

for all creatures small and strange

contents

brown pelican

1

Familiars

I might have been the fish the brown pelican swallowed.

—Jane Hirshfield, "Day Beginning with Seeing the International Space Station and a Full Moon Over the Gulf of Mexico and All Its Invisible Fishes"

Walter Inglis Anderson loaded his supplies into his small, hardly seaworthy skiff: a small ream of cheap typewriter paper, a notebook, pens, pencils, watercolors, and enough sepia ink to last a week or three. He was a New Orleans–born bohemian, an eccentric father of four who required solitude, a schizophrenic who escaped his last stay in a mental hospital by knotting bedsheets and lowering himself down from an upper-story window. Following his escape, orderlies discovered that he had filled the walls of his room with soap-drawn sketches of birds, aflight and free.

Anderson was an artist—a prodigious and, in his lifetime, sorely unappreciated one. Following six years of art school in New York

Opposite: Close-up of the stately brown pelican. Photograph by Grayson Smith, U.S. Fish and Wildlife Service.

and Philadelphia, he moved to Ocean Springs, Mississippi, where he assisted in his brother Peter's pottery studio. There, he began to perfect the singular, meditative vision that would define his life's work: tracing the mystic chords that connect humanity to the natural world, or what he called "the miracle of realizing that art and nature are literally one thing."

Beginning in the mid-1940s, after turning forty and shortly following his last diagnosed psychotic break, Anderson started rowing thirty miles out to the Chandeleur Islands, off the coast of Louisiana, to find peace, to create art. He bivouacked on North Island, which he called North Key, high atop an oyster shell midden, warming himself by a driftwood fire, amid a nest of mangroves, in the midst of the world's largest colony of brown pelicans.

"The whole island seems a concentration of life," he scribbled in his notebook: shorebirds and migratory passerines, seabirds and sea life, mosquitoes and no-see-ums. He captured this island menagerie—his "familiars," he called them—with ink-dipped crow quills, paper, and paint. Birds were his main inspiration: ducks, herons, and egrets, terns and gulls, red-winged blackbirds, storm-borne frigates, ospreys on the hunt and pipers at play, cormorants, mergansers, grackles, gallinules. According to his daughter Mary, "When he saw a bird, he was that bird." But he cherished the brown pelican most of all. An inescapable presence along the Gulf Coast, pelicans were his obsession, his most familiar familiar, "his totem and a spiritual guide."

Who knows why we pick a person, a plant, an animal to fall in love with. For Anderson, perhaps it was fate—he was born the same year,

1903, that President Theodore Roosevelt saved the brown pelican from local extinction by designating the nation's first wildlife refuge at Pelican Island, Florida. The following year, Roosevelt carved the nation's second wildlife refuge from the Chandeleurs and the adjoining Breton Islands. Shuffling such synchronicities aside, we know that Anderson took pleasure in being in the presence of pelicans and never tired of portraying them on the page. "He might draw a hundred pelicans," one biographer wrote, "or one pelican a hundred times." And on North Island there were many thousands of pelicans to draw.

Anderson sketched the pelicans in what he called "all their reactions and conditions of life": soaring, feeding, returning, repeating, loafing, nuptialing, nesting. He reveled at their synchronized movements, their "tremendously musical harmonies of rising from the ground at my approach." He learned to live not only with pelicans but as pelicans live: bathing as they bathed, walking the beaches barefoot just like they did (though he lacked their four webbed toes). He slept in a makeshift tent or under his boat, but sometimes gathered dried grass, he wrote, "to make myself a nest like the Pelicans." He even claimed to have learned the pelican's language, and compiled a lexicon, the "Pelican Dictionary of Common Terms," despite the fact that the bird, once it has fledged and left the nest, hardly ever speaks:

Tchoo: falsetto, used endearingly.

Arp or *arp bow wow:* [no definition provided.]

Arrh-h-h: when the mother flies over, perhaps with food.

Why-ah: very shrill.

Kerr-uph: [no definition provided.]

Arr-uh-Arr-uh: when the mother bird arrives with food.

Quieh-hh-qui-eh-h: often accompanied by flapping of wings.

Aw-awh-aw-aw: [no definition provided.]

Pelicans were the artist's agony and his ecstasy. He mourned the sick, buried the dead, and celebrated the pelican colony's rebirth each spring. He subsisted on fish and oysters, but often ran out of food and resorted to salvaged tin-can jetsam, their labels worn away, or the rare pelican egg, rolled free from its nest. They "taste good but the consistency [is] nauseating," he admitted, and are "better raw than cooked." Anderson consumed pelicans and they consumed him, literally. One afternoon, while drawing, he was "attacked several times by a pugnacious young pelican who tried to swallow [his] bare foot." A few days later he took part in an "ecstasy of feeding" of hungry nestlings. "Today I assisted at an orgy," he wrote, "a religious . . . feast with the passion and excitement of sexual delirium. I drew while it went on, sitting beneath the mangrove bushes."

"After you have lived on the island for a while," Anderson wrote, "there comes a time when you realize that the pelican holds everything for you." He resembled the weary but resolute wanderer featured in Psalm 102: "I am like a pelican of the wilderness: I am like an owl of the desert. / I watch, and am as a sparrow alone upon the house top." Alone, he never grew lonely. But despite the countless hours he spent in their presence, for all the times he put pelican to page, the artist regretted that "it's impossible to get all you want." There were always more pelicans to paint, until the time came when there weren't.

* * *

Unlike Walter Anderson, I have long been frightened of birds. Blame Norman Bates, the pet cockatiel that moved in with my stepfather when I was six years old. Norman Bates, you might have guessed, was named for Alfred Hitchcock's sociopathic antihero. The two, my stepfather warned, shared a taste for bloodshed. I mostly stayed clear of his cage, wedged in the corner of our tight-squeezed dining room, where dinner meant facing my fear of losing an ear. On the rare instance when my stepfather—despite my mother's protestations— allowed Norman Bates some unrestrained, in-house exercise, I would scramble for the safety of my bedroom's closed door. Even today, nearly twenty years after Norman Bates's death, I cannot watch *Psycho*'s iconic shower scene without imagining that cute little bird, with those bright orange cockatiel cheeks resembling twin splashes of blood, slashing the curtain with his claws.

Birds, besides Norman Bates, were for me both there and not there, omnipresent and invisible. Though I grew up surrounded by water and trees—prime bird real estate—I paid those flying, twittering creatures no mind. Until I became a passionate, day-to-day, list-building birder during the coronavirus pandemic, I could identify only a dozen or two birds common to my home state of Louisiana, including the brown pelican, of course. She was that plump-pouched beggar pleading for scraps near fishing docks, the beachfront drifter, my state bird, emblazoned so beautifully—not to mention beatifically—on our state flag, which, with an image of a Christlike mother pelican piercing its breast to nurture her chicks on her own lifeblood, should come with a trigger warning for anyone with a blood phobia, a complicated relationship with the Catholic Church, or a mother.

But what most defined the pelican for me was the summer blockbuster that arrived in theaters one month before I turned thirteen: *Jurassic Park*. Allow me to narrate the film's final minutes, in case you've forgotten or, perish the thought, have never seen Steven Spielberg's masterpiece. A tyrannosaurus rex does dino battle with a pair of raptors, allowing the failed theme park's survivors to escape Isla Nublar by helicopter. As John Williams's score soars toward the credits, the camera pans around the chopper's cabin. John Hammond, played by Richard Attenborough, remorsefully regards his amber-encased mosquito, while his two grandkids nestle in the arms of Sam Neill's Dr. Alan Grant, who finally accepts his destiny to be a father. Laura Dern's Dr. Ellie Sattler sits across from him, expressing her love with tear-brimmed eyes. She glances outside; he follows her gaze. Five brown pelicans flap their wings, majestically gliding into a Pacific Ocean sunset. That brief but remarkable montage distills and dramatizes the film's main message: that "life finds a way"—humans will make babies, humans will make mistakes to the great detriment of those babies, dinosaurs will evolve into birds, and the cycle of *Jurassic Park* sequels will never end.

Whoever picked the pelican from the casting call's list of Hollywood hopefuls keenly chose the right bird. The earliest known pelican fossil, excavated from the Luberon mountains of Southern France, dates to the early Oligocene Epoch—some thirty million years ago. That proto-pelican is nearly identical to today's bird, though endowed with a slightly longer beak. For that reason, the brown pelican and her seven sister species of the family Pelecanus—the great white pelican, Peruvian pelican, American white pelican, pink-backed pelican, spot-billed pelican, Australian pelican, and Dalmatian pelican—have been called "living fossils" and "one of the most superficially dinosaur-like of the world's 10,000 or so bird species." Walter Anderson agreed, grouping pelicans with the "earlier and more primitive forms of life" and dubbing them "prehistoric monsters."

* * *

Anderson stopped visiting North Island at some point in the late 1950s, switching his allegiance to Mississippi's Horn Island. Maybe the aging artist found the shorter, but no less arduous, trek to Horn more feasible. Perhaps a recent string of hurricanes had finally rendered North Island's silty shores uninhabitable. But I like to think he stopped rowing out to the Chandeleurs once the pelican population plummeted.

He speculated—correctly, it would turn out—that pesticides were to blame for the eclipse of the brown pelican along the Gulf Coast, but biologists would not validate that hypothesis until several years after his death. All Anderson could do was watch and paint as the pelican vanished—first gradually, then dramatically—and was finally snuffed out completely, like a beachfront campfire in a tempest. "In a word," he would write about the bird, "you lose your heart to it. It becomes your child and the hope and future of the world depend on it." The pelican had disappeared from North Island, and with it had gone the artist's hope, his future, his world.

In mid-June 1965, Anderson readied his skiff for what would become his penultimate island adventure. On his second day, he walked to Horn Island's far westerly point, where, a mile in the distance, he spotted a pair of pelicans sailing out to sea. He celebrated his good luck. "Providence," he wrote, "has been kind." Eleven days later, fortune smiled upon him once again: "Three pelicans—then six pelicans," he rejoiced. "Time has gone back." Independence Day greeted him with a miracle: the largest pod of pelicans he'd witnessed in well over a decade. "Pelicans!" he wrote, "seventeen in one flock!" On September 9, Hurricane Betsy roared ashore, decimating Horn Island, where Anderson rode out the storm with his skiff roped to his waist. As the sun greeted the next morning, he discovered his island of familiars gone. Less than three months later, on November 30, Walter Anderson would succumb to lung cancer in a New Orleans hospital.

That same month, President Lyndon B. Johnson's Science Advisory Committee released the nation's first comprehensive report on what we would, decades later, understand to be a global climate crisis. "Pollution now is one of the most pervasive problems of our society," the report warned. "Vast quantities of wastes and spent products . . . pollute our air, poison our water, and even impair our ability to feed ourselves." By decade's end the West Coast's largest brown pelican colony would, like the one on North Island, dissolve due to exactly what that report heralded. The living fossil risked fossilization, not for the first time and not for the last, as man and bird both stumbled toward extinction.

This book tells the story of our fraught and often savage relationship with a single species of bird, the brown pelican (*Pelecanus occidentalis*), but inevitably it also speaks to our relationship with the planet and our relationship with ourselves, our own humanity, or lack thereof. In the roughly 150 years of history this book traces, the pelican has been poisoned, beaten, mutilated, and murdered. It's worth remembering, bears repeating like a secular mantra: this did not have to happen. No part of this had to happen.

For many millennia, the pelican has existed as a symbol of saintly piety. Today, the brown pelican is considered a bellwether bird, or, for biologists, an indicator species, able to portend what disasters await. In truth, those two metaphoric pelicans—one mythic, the other scientific—are one and the same. If "'hope' is the thing with feathers," as Emily Dickinson famously wrote, without feathers, we can presume, hope dies.

The brown pelican provided Walter Anderson with hope and with a purpose—several purposes, perhaps: artistic, spiritual, communal. For a man who found it hard to love his fellow man, Anderson found in pelicans a creature to love. But love, as he well knew, as we all know, equals loss. The grief he must have felt as his most precious familiar became unfamiliar, is a grief—a grief for our planet, its creatures, ourselves—we will all share as we head into an uncertain future. That much is certain.

2

Pelicanland

For every bird there is a stone thrown at a bird.

—Maggie Smith, "Good Bones"

Over a pair of late afternoons in early February 1886, a promising twenty-one-year-old ornithologist named Frank Chapman strolled the sidewalks of Manhattan's 14th Street fashion district counting birds. He tallied five blue jays, nine Baltimore orioles, twenty-three cedar waxwings, and sixteen quail. He spotted a pileated woodpecker and a scissor-tailed flycatcher, dissonant sights among the city's hustle-bustle. But undoubtedly his strangest observation of all was a greater prairie chicken, a bizarre-looking bird whose habitat doesn't stretch far beyond the grassy plains of the Midwest.

Chapman identified forty species, 173 birds in total, and not one was found perched in a tree, bathing in a fountain, or on the wing, navigating New York's pre-skyscraped heavens. Instead, each and every

Opposite: Warden Paul Kroegel at Pelican Island National Wildlife Refuge. Photographer unknown, U.S. Fish and Wildlife Service.

bird on Chapman's list decorated a woman's hat. Beginning in the late Victorian era, feather fashion was all the rage. Voguish urbanites on both sides of the Atlantic wore extravagant, wide-brimmed hats festooned with silk flowers and ribbons and flocked with feathers: the snowy egret's downy frills, the neon-pink plumes of a roseate spoonbill, the bald eagle's power quills. Enterprising milliners, however, did not settle for simply sticking feathers in a cap. Entire nature tableaux spilled forth from ladies' chapeaus: fruits and flowers, grasses and mosses; triumphant scaffoldings of emu and peacock quills; a pair of severed wings in flight; an owl's head; a bouquet of dainty warblers; an entire hummingbird, greater bird of paradise, or some other poor, taxidermied avian soul.

Up and down the Atlantic and Gulf coasts, hired huntsmen engaged in the wholesale butchery of birds. At the carnage's height, an estimated five million were killed annually, mostly in Florida, leading to the widespread migration of waterfowl farther south in search of safe nesting sites and the near extinction of several species. Louisiana, according to one local ornithologist, ranked as a "slaughter ground of the plume and wing hunters." Reporters called it a Plume War and the Age of Extermination.

In swooped George Bird Grinnell, whose felicitous name, not to mention biographical bona fides, would help establish him as the plume trade's public enemy number one. While a paleontology graduate student at Yale, Grinnell joined George Custer's 1874 colonial expedition into the Dakotan Black Hills as a naturalist. (He luckily declined the cavalry officer's follow-up invitation to Little Big Horn.) The next year he explored the newly established Yellowstone National Park and, horrified by the widespread bison poaching there, published articles and lobbied presidents on the critically endangered animal's behalf. As editor of the popular outdoors magazine *Forest and Stream,* Grinnell published Chapman's original catalog-as-clarion-call, titled "Birds and Bonnets," just two weeks after his own announcement instituting the nation's first Audubon Society.

In 1886, John James Audubon had been dead for three and a half decades, but for the country's close-knit community of bird enthusiasts, it felt as if America's most notable naturalist still stalked its fields, woods, and waters in search of subjects to paint. "The land which produced Audubon," Grinnell wrote in a front-page *Forest and Stream* editorial, "will not willingly see the beautiful forms he loved so well exterminated." The Audubon Society, Grinnell declared, would endeavor to protect all wild birds—except those killed for food—and their eggs.

To introduce his project, Grinnell did not appeal to the better nature of milliners or plume hunters, but rather indicted the same people who might, or might not, choose to wear feathered hats: what one of his closest partisans called the "dead bird wearing gender."

"The reform in America, as elsewhere," Grinnell wrote, "must be inaugurated by women, and if the subject is properly called to their notice, their tender hearts will be quick to respond." From that first New York chapter, Audubon societies soon sprung up nationwide and their influence became evident: hat makers shifted from incorporating birds to shiny beads; women tossed off their bonnets; and children learned to hunt sans shotguns. Emboldened, Grinnell launched the *Audubon Magazine* the following year, featuring the bylines of newly enlightened hunters—"How I Learned to Love and Not to Kill" was the title of one article—and numerous women, who, more often than not, were obligated to remain on their side of the gender divide. "We flattered ourselves that the tender and compassionate heart of woman would at once respond to the appeal for mercy," wrote Celia Thaxter in an essay titled "Woman's Heartlessness," "but after many months of effort we are obliged to acknowledge ourselves mistaken in our estimate of that universal compassion, that tender heart in which we believed."

The Audubon Society and its magazine folded as quickly as they had appeared, chiefly for the fact that Grinnell was a hypocrite. For men like Grinnell, bird conservation often meant bird and egg collecting, which meant bird killing. This was the era of "shotgun ornithology," when birdshot, not binoculars, brought man closer to birds; the era when oology, or the study of bird eggs, meant plundering nests for as many unhatchedlings as one could carry. *Forest and Stream,* which Grinnell headed until 1911, was, of course, a fishing and hunting magazine, which carried the subtitle, "A Weekly Journal of the Rod and Gun." If anything, Grinnell was continuing in the tradition of his hero, his bird-saving society's dedicatee, who famously killed many more birds than he could ever hope to paint. If we could poll a sample of voting-age birds, they'd no doubt elect John James Audubon exterminator in chief. Reading Audubon's voluminous writings today can be an exercise in cognitive dissonance between savoring the naturalist's frequently gorgeous prose and wishing to reach back through time and strangle the man for his callous disregard for wildlife. If you've never read Audubon, please allow me to quote at length an excerpt from his December 31, 1831, excursion south of St. Augustine, Florida:

I was anxious to kill some 25 brown Pelicans to enable me to make a new drawing of an adult male bird . . . I shot some rare birds, and putting along the shore, passed a point, when lo! I came in sight of several hundred pelicans perched on the branches of mangrove trees, seated in comfortable harmony . . . I waded to the shore under cover of the rushes along it, saw the pelicans fast asleep, examined their countenances and department well and leisurely, and after all, leveled, fired my piece, and dropped two of the finest specimens I ever saw. I really believe I would have shot one hundred of these reverend sirs, had not a mistake taken place in the reloading of my gun. A mistake, however, did take place, and to my utmost disappointment, I saw each pelican, old and young, leave his perch and take to wing; soaring off, well pleased, I dare say, at making so good an escape from so dangerous a foe.

Even a bleeding heart like Frank Chapman could succumb to the era's collection-as-conservation ethos. In 1889, on a field exposition along the Atlantic coast of central Florida to gather specimens for the American Museum of Natural History, he killed fifteen endangered Carolina parakeets, an act he immediately regretted. "Good luck to you poor doomed creatures," he wrote in his journal, "may you live to see many generations of your kind." Chapman would outlive the species, which went extinct within two decades.

Virginia Woolf perhaps understood the situation best. In a 1920 essay, the British writer invited readers to weigh the sins of the plume purveyor versus the feathered-bonnet bearer. Reconsider that "lady of the stupid face and beautiful figure . . . going tonight to the opera," Woolf wrote. "Plumes seem to be the natural adornment of spirited and fastidious life, the very symbols of pride and distinction. . . . A lemon-coloured egret is precisely what she wants." Woolf gathered steam: "We may fairly suppose then that the birds are killed by men, starved by men, and tortured by men—not vicariously, but with their own hands." And then, she dropped her final indictment: "Can it be that it is a graver sin to be unjust to women than to torture birds?"

The National Audubon Society we know today emerged in 1896, a decade after Grinnell's first, failed iteration, when Boston socialite cousins Harriet Hemenway and Minna B. Hall, sickened by tales of blood-soaked egret rookeries, resolved to revive the war against plumes. The cousins gathered upstanding lady friends over tea, pledged to boycott bird hats, allowed a handful of men to join the group, and together formally created the Massachusetts Audubon Society. This new, female-led, grassroots reform movement quickly spread across the country. Early on, women accounted for 80 percent of membership rolls and many executive committee positions. (This mostly remains the case today. At the time of writing, my local New Orleans Audubon chapter is entirely led by women.)

The Audubonites educated, lobbied, and preached. They circulated lists of "quills to avoid" and encouraged good citizens to trade firearms for field glasses (today's binoculars). They helped pass pro-bird, anti-plumage federal legislation, most significantly the Lacey Act of 1900, which banned trafficking in illegal wildlife, and the 1918 Migratory Bird Treaty Act, what the National Audubon Society calls the nation's "best bird law" (and one that was cruelly gutted in the final days of Donald Trump's administration in deference to the petrochemical industry).

But for every feathered hat relegated to refuse, a rising middle class ensured that two would replace it. "That there should be an owl or ostrich left with a single feather apiece hardly seems possible," *Harper's Bazaar* proclaimed during the winter hat season of 1897. Trends die hard. The plume war recommenced. And a new feather rose in popularity. A rich reddish brown with white-to-silvery-gray highlights—a bit drab, perhaps, but plentiful, easy to procure, and able to be paired with anything—the brown pelican's plumes had arrived.

"The feathers of this bird are now worn so commonly—hundreds may be seen in New York City daily," Frank Chapman wrote in an 1899 editorial, "that at the present rate of destruction[,] its early extinction, at least in the United States, is assured."

* * *

It's worth pausing here to note that the brown pelican had long been seen as a throwaway bird at best, and at worst, as we shall see, a pest ripe for eradication. The ways in which writers have described the bird over time range from desultory to despicable. May I present a small sample: Pelicans are "lazy." They are "heavy and clumsy." "Homely." "Graceless." "Gluttonous." "Gawky." "Ludicrous." "Queer." "Strange and comical." "Strange, weird creatures." "Such peculiar birds." "Awkward, ungainly birds." "Silent, dignified and stupid birds." "This tragicomic bird." "This comedian of the seas." "Extremely stupid." "Stately and solemn." "Stately but slow-witted." The bearer of a "sad melancholy expression."

"Pelicans," according to one notable ornithologist of old, "are not emotional birds."

Compared to its cousin, the snow-downed, swan-like spectacle known as the American white pelican, the brown pelican is a pariah. "I know of no more magnificent sight," wrote eminent birder Arthur Cleveland Bent, "than a large flock of white pelicans in flight." On the other hand, the brown pelican, in Bent's estimation, is "grotesque."

Others agree. Brown pelicans are "incongruous and grotesque." "Grotesque fishermen," with "grotesque and striking features." "The least attractive from an esthetic point of view." "The ugliest and most awkward bird we have." "A disagreeable, wheezing, asthmatic bird which would take as much pleasure in plunging its hooked beak into the eye of friend as foe."

These descriptions, it should be noted, all come from authors writing *in praise* of the pelican. Their point is clear: beautiful this bird is not.

* * *

Frank Chapman had come a long way since tallying plumed hats. In 1888, he joined the American Museum of Natural History's ornithology department, and over the next dozen years he traveled some 90,000 miles, documenting bird habitats and collecting specimens. By century's turn, the museum named Chapman their first curator of birds, responsible for designing an exhibit hall dedicated to the continent's feathered populace.

Before the grizzly bear family and the chest-pounding mountain gorilla; before the *Australopithecus* couple tentatively stepping toward us, his woolly arm casually slung over her hairy shoulder; before the squid and that great blue leviathan, the whale, the Hall of North American Birds was the museum's apotheosis of nature captured in situ, transplanted and installed on Manhattan's Upper West Side, and through the magic of art and taxidermy brought back to life.

When Chapman's Hall of North American Birds opened in 1902, its thirty-four dioramas combined to form the world's first museum space entirely devoted to modeling large-scale, three-dimensional, painted habitats. Its centerpiece revealed a tiny, triangular island located off Florida's Atlantic coast, smack dab in the middle of the Indian River Lagoon, the continent's most biodiverse estuary, a brown pelican paradise spoken of in hush, reverential tones and visited by only the most hardcore birders, a place Chapman had explored once before, a site he called "Pelicanland."

President Theodore Roosevelt toured the hall and immediately recognized the Pelican Island diorama to be a masterpiece. The display fulfilled the mission of his father, who cofounded the museum in 1869: to employ the natural sciences to enlighten, inspire, and maybe even entertain. In the diorama's foreground, a baker's dozen of taxidermied pelicans showcased the life stages of the species: adults brood on an immense mattress of mangrove branches, a pair of juveniles playfully butt heads, twin chicks stretch their giraffine necks down their parent's gullet to feed on regurgitated fish. Dozens, if not hundreds, of pelicans fill the painted background. The diorama, which is no longer on display after being dismantled and disposed of several decades ago, was nothing short of breathtaking.

All of this is under threat, Chapman told his friend Roosevelt: *this five-and-a-half acre Pelicanland—that you, the federal government, own—will soon be annihilated by the plume trade. You can help, Mr. President. You can save Pelican Island.*

The president was famously a roughriding outdoorsman, big game hunter and sport angler, wannabe cowboy, and, perhaps most surprisingly, a passionate birder. For a man who likened himself to a bull moose and an old lion, it might come as a surprise that Roosevelt carried a soft spot for the more tender species of the avian kingdom. He cherished what he called the "rich melody," the "wild sadness" of the meadowlark's cry, the "gentle, hopeless" coo of the mourning dove. Wherever he traveled—touring Europe after winning the Nobel Peace Prize, on safari in Africa—he kept careful notes on what birds he heard and saw. He compiled a life list—fifty-five species in total!—of birds spotted on the White House grounds, and wrote ornithological academic articles with titles like "Revealing and Concealing Coloration in Birds and Mammals." He distrusted any foe of fowl, and once left Yosemite positively perplexed that his friend John Muir cared not a lick for birds, grieving that "the hermit thrushes meant nothing to him."

Roosevelt, in contrast, equated a sustainable bird population with a healthy home, a robust America. And though he hunted wildfowl, he detested the indiscriminate butchering of any bird. "I need hardly say how heartily I sympathize with the purposes of the Audubon Society," Roosevelt wrote Chapman in 1899.

I would like to see all harmless wild things, but especially all birds, protected in every way. Spring would not be spring without bird songs, any more than it would be spring without buds and flowers, and I only wish that besides protecting the songsters, the birds of the grove, the orchard, the garden and the meadow, we could also protect the birds of the sea shore and of the wilderness. . . . When the bluebirds were so nearly destroyed by the severe winter a few seasons ago, the loss was like the loss of an old friend, or at least like the burning down of a familiar and dearly loved house.

He went on to lament the rapid disappearance of the passenger pigeon and Carolina parakeet, two recently extinct North American avifauna. "When I hear of the destruction of a species," he concluded, "I feel just as if all the works of some great writer had perished."

The brown pelican might not have been Roosevelt's gateway bird, or even his favorite, but he no doubt prized its company. He would have been familiar with the species from time spent on the beaches of Florida and Cuba during the Spanish-American War. "To me they are always interesting and amusing birds," he would write many years later, "and I never get over the feeling of the unexpected about them."

For a budding conservationist in tune with threatened birdlife and in touch with reformed bird lovers, Pelican Island provided an opportunity. Roosevelt understood the consequences of inaction. Although he had never visited the island, he knew of its history, its lore and beauty. The area likely first entered his ornithological orbit through his beloved "Uncle Rob," Robert Roosevelt, a conservationist-minded sportsman, progressive politician, and wide-ranging writer—everything young Teedie, as his family called him, hoped to become. Uncle Rob adored the Indian River—he cherished casting lines in the lagoon's brackish waters and treasured the area's citrus bounty. "Remain there," he encouraged visitors, "till your heart is glutted with sport, and your palate with fruit, and thence return to the North by rail or boat. Such a trip makes a date of delight in one's life."

The region captured Roosevelt's imagination for a second time in college, while he was perusing the field notes of fellow Harvard man Henry Bryant. Decades earlier, Bryant spurned a promising career in surgery to dedicate his life to birds. As curator of ornithology for the Boston Society of Natural History, he led excursions to gather bird skins and eggs across Europe and the Americas. In 1858, while sailing up the Indian River, counting and collecting birds, he marveled at the number of pelicans. "I found them breeding in larger and larger numbers as I went north," he wrote, "until I arrived at Indian River, where I found the most extensive breeding-place that I visited . . . Pelican Island." The island sustained a population of over 10,000 pelicans, in addition to untold thousands of nesting egrets, herons, and spoonbills, as well as visiting frigates and ibises, all nestled atop the island's thick mangrove forest, shrubs the size and shape, in Bryant's description, of "large apple-trees."

Word spread of Bryant's cacophonous, motley-feathered rookery. And in 1879 an ardent fisherman and travel writer named Dr. James Henshall skippered a skiff to Pelican Island and found a decidedly less idyllic scene. "The mangroves and water-oaks of this island have been all killed by the excrement of the pelicans," he wrote in *Camping and Cruising in Florida.* "This guano, which lies several inches deep on the ground, is utilized by the settlers as an efficient fertilizer. At a distance the dead trees and bushes and ground seemed covered with frost or snow, and thousands of brown pelicans were seen flying and swimming around or perched upon the dead branches."

Before Henshall could get any closer, a flurry of shotgun blasts churned the sky into a scattering of wings, sending him into a panic. "We saw a party of northern tourists at the island, shooting down the harmless birds by scores through mere wantonness. As volley after volley came booming over the water, we felt quite disgusted at the useless slaughter, and bore away as soon as possible."

Henshall's report was doubly troubling. Pelican Island appeared to be a site of convergence, a sanctuary, the last breeding colony for brown pelicans on Florida's Atlantic coast. Their large numbers not only overburdened the island's fragile ecosystem but became an easy mark for plume hunters or those simply looking for some target practice. Visiting the island over a decade later, the oologist Morris Gibbs felt surprise that the pelicans "adhere to a chosen site even when persecuted year after year." Though local residents informed him that the hatchling population numbered ten to twenty times higher just a few seasons back, he carried off nearly two hundred eggs for his personal collection and to trade with friends.

Alarmed by such reports, Frank Chapman decided to survey the now legendary island in March 1898. To describe this bird nerd as anxious would be a vast understatement. "No traveler ever entered the gates of a foreign city with greater expectancy," he said, "than I felt as I stepped from my boat on the muddy edge of the City of the Pelicans." We might also attribute Chapman's nerves to the fact that he was accompanied by his new wife, Fanny. This was their honeymoon. Over four days, the Chapmans counted occupied nests, fledglings, and adults, crunched the numbers, and estimated just over 2,700 pelicans. On a follow-up visit two years later, he discovered a population drop of almost 15 percent. Not only had the island's inhabitants been, in his words, "brutally persecuted" by plume traders, but sloops and yachts had begun to arrive daily, depositing tourists during the height of the February–March nesting season. Brooding pelicans fled at the first sight of these unwelcome day-trippers, leaving their young and unhatched eggs to the mercy of the sun's scorching rays.

Despite the setbacks, Chapman remained buoyant, deeming his Pelicanopolis "by far the most fascinating place it has ever been my fortune to see in the world of birds." But he also departed with a warning. "If the natives of the state ever open their eyes to the indisputable fact that a living bird is of incalculable greater value to them than a dead one," he wrote, "they may perhaps take some steps to defend their rights, and by passing and enforcing proper laws, put an end to the devastations of the northern plume agents, who have robbed their state of one of its greatest charms."

By March 1903, Pelican Island's fate rested in President Roosevelt's hands. He had read Chapman's reports and knew that the American Ornithologists' Union (AOU)—a slightly older and less political contemporary of the National Audubon Society—had tried to purchase the island from the federal government. That plan took a cue from Edward Avery McIlhenny's private Louisiana wildlife refuge, Bird City, established on the Tabasco heir's Avery Island in 1895. Within two decades, McIlhenny nurtured eight snowy egrets into a population numbering in the many thousands, saving the species from extirpation, or local extinction. Chapman and AOU official William Dutcher called a White House meeting on March 14, 1903, to plead their case for intervention. "Is there any law that will prevent me from declaring Pelican Island a Federal Bird Reservation?" Roosevelt barked to an aide. Unbeknownst to the president, Chapman and Dutcher had prearranged for the Department of Agriculture to sign off on the deal. *No,* came the response, *declare as you wish, Mr. President.* "Very well then," Roosevelt trumpeted, in what would become a catchphrase of sorts, "I so declare it."

"To see a file of pelicans winging their way home across the afterglow of sunset, is like a gallery of old masters," Roosevelt would say a decade later while standing on the shores of the Breton National Wildlife Refuge. "Their loss would be like the loss of artists for all time."

* * *

By executive order, the nation's conservationist in chief had just established the first federal bird reserve. But laws do not enforce themselves. Pelican Island needed a protector, a wildlife warden to keep poachers and tourists at bay. Luckily, Paul Kroegel, favorably known as the "Audubon of the Indian River" in birding circles, already patrolled the waters surrounding the island.

Born in Chemnitz, Germany, Kroegel immigrated with his father to the United States in 1870, at the age of six. After short stints in New York and Chicago, father and son headed south, seeking sunshine and citrus, eventually homesteading on an oyster-shell Indian mound overlooking Pelican Island's lagoon. They lived in a palm-thatched hut, grew grapefruit and oranges, and tended a bustling beehive colony. Kroegel learned to build boats—from workaday canoe to recreational schooner—and to negotiate the river's dense mangrove swamps. By his teenage years, locals had nicknamed him the "Pelican Watcher." He couldn't help but love and perhaps identify with the odd birds. Kroegel wore a droopy mustache—twin wings that flapped over his omnipresent tobacco pipe—and heavy wool sweaters impervious to a mosquito's bite. A loner, maybe even a misanthrope, he married and raised kids, but pelicans were his one true brood. He slept with one eye open, according to family lore, to monitor for any strange ships—potential plumers—passing in the night. At the sound of an anchor's splash, he would grab his double-barreled ten-gauge, jump in his little yellow sailboat, named *Yellow Kid,* and do his damnedest to save some birds. Beginning in 1902, the Florida Audubon Society paid him a small stipend. Now he earned a government paycheck of fifteen dollars a month and came armed with a federal badge: Pelican Island Warden.

There was only one problem, a monumental hiccup that quickly became apparent during the new national wildlife refuge's first official nesting season: all the pelicans of Pelican Island had disappeared. Audubonites blamed the massive no trespassing sign, visible clear across the Indian River, that read "U.S. RESERVATION—KEEP OFF." Kroegel replaced the eyesore with a significantly smaller sign, and the pelicans returned the following year. "Consequently," Chapman later joked, "we may infer from this incident either that the Pelican can read and has strong political prejudices which prompt it to refuse favors from the administration which has preserved its home, or that it lacks sufficient discrimination to realize that a board painted white with black marks and held upright by two posts is perfectly harmless."

Once the pelican population returned, the Pelican Watcher settled in to observe "all the activities of Pelican home-life"; as Chapman recorded after a morning spent in the warden's company, these included "nest-building, laying, incubating, feeding and brooding young, bathing, preening, yawning—and the Pelican yawn is indeed a yawn—sleeping." Kroegel more than earned his monthly fifteen-dollar paycheck, filing regular reports from the island. He studied how male pelicans choose their breeding sites in late autumn, observing their performance of a head-swaying display to attract a potential mate. Together, the new couple construct their nest. Over one week, he gathers, she builds. Trees make for prime real estate, a safe haven from the floodwaters that frequently wash over the island from the east. Stragglers make do with ground nests, stacking mangrove sticks, straw, grass, and leaves higher, ever higher, sometimes stretching three feet tall. Copulation lasts fewer than ten seconds but may be repeated several times within an hour. Eggs—two to three per clutch—appear in November or December, after which both parents share incubation duties, switching off in a performative dance Chapman called the "Ceremony of Nest Relief."

Warden Kroegel's duties were not without danger. Plume and egg poachers still canvassed the reserve, and he arrested many over the years, allegedly including the robber baron Andrew Mellon. His jurisdiction, however, only covered the small island and its encompassing waters. Elsewhere, game wardens, as well as birds, could not rely on the security of federal protection. In July 1905, outside the town of Flamingo, Florida, plumers murdered Guy Bradley, an Audubon Society warden who patrolled the Everglades and Keys. Three years later, Audubon warden Columbus G. McLeod disappeared while tracking egret hunters near Cayo Pelau Island, off Florida's Gulf Coast. Authorities found McLeod's sandbagged boat sunk in the bay and his bloodstained hat, but never located the warden's body. Their two deaths incited Congress to finally pass the Migratory Bird Act of 1918.

But that same year, poachers found a new reason to wage war against the pelican. Fishermen, especially those who plied their trade along the Gulf Coast, blamed the bird for depleting their stock, for interrupting their income stream. They accused brown pelicans of eating upwards of a million dollars' worth of fish and seafood destined for human consumption—mullet, mackerel, trout, pompano, shrimp—daily. Millions of pelicans haunt the waters of the American South, they said, hungry Hitchcockian hordes scouring the sea for sustenance.

Humanity had long feared, erroneously, that brown pelicans survived on what we think of as "food fish." "It has been a question with me for some years," pondered egg baron Morris Gibbs, "whether many birds were not enemies to man by reason of their destruction of certain animals and vegetables upon which man depends." Consider the pelican, he offered, and "its vast destructive powers upon the fishes, wherever it is found." He guesstimated some numbers (correctly, as it would turn out): an adult pelican consumes around sixteen fish, or four pounds, per day; thus a colony of five thousand birds will require well over seven million pounds annually. "On principle," he concluded, "it might be well to kill off these birds."

It didn't help that in 1918, all sectors of the nation's economy, including industrial fishing, shifted into overdrive to supply the war effort. "The claim is being put forth," the National Audubon Society's official organ decried, "that Pelicans are eating up the fish at such a rate that the birds must be destroyed if we are going to have sufficient food to feed our people and win the war."

"Kill the pelican," went the anti-pelican battle cry, "or the Kaiser will get you."

Fishermen found a friend in the press. A bellicose editorial in the *St. Petersburg Evening Independent* argued that

it is time that the Government was informed as to the destruction that is being wrought by Pelicans in southern waters. The Pelicans are protected by a National law and therefore are thriving and increasing in number, and it will be only a few years until the people will have to choose between the Pelicans and the fish. The Pelican is no earthly use to anybody and serves no useful purpose. The fish are needed to help supply the deficiency in food. One Pelican will consume 100 to 300 small fish in a day. Multiply that by the thousands of Pelicans in this section, and you have some idea what the Pelicans do to destroy fish. It is a tremendous price the Government is paying to satisfy a few sentimentalists who want to save the birds.

Calls rang out to close all state and federal bird reserves. The deputy fish commissioner of St. Petersburg advanced a proposal to destroy all pelican eggs for a period of five years, a plan that the *Evening Independent* heartily endorsed. Down the coast, the Texas legislature reviewed, and ultimately failed to pass, two bills that authorized the liquidation of pelicans and gulls. In New Orleans, at a meeting of fish and food commissioners from the five Gulf states, attendees introduced a motion to exterminate all birds that eat fish. A top administrator in the federal agency responsible for feeding soldiers fighting overseas publicly stated that "the Pelican serves no useful purpose whatever."

Unsurprisingly, the brown pelican and her seafaring sistren were not alone in attracting the ire of bloodthirsty Americans. Legislative bodies targeted other avian "pests" nationwide: cherry-snatching robins in New Jersey, alfalfa-hungry mourning doves in Arizona, meadowlarks accused of decimating California's grape harvest. In April 1917, the Alaska Territorial Legislature issued a bounty of fifty cents for each pair of bald eagle talons delivered, resulting in the liquidation of over five thousand birds. In the 1920s, Yellowstone National Park rangers killed hundreds of the breeding white pelican population to satisfy sport anglers' complaints that the bird had caused the decline of the cutthroat trout.

Florida's anti-pelican constituency would not wait for legislative approval. "A new carnival of bird slaughter" commenced, in the words of the state's Audubon society. On the night of May 10, 1918, a group of assailants landed on Pelican Island and clubbed four hundred pelican nestlings to death. Even a national wildlife reserve remained unsafe. Kroegel's warden appointment had been gradually phased out, a casualty of wartime thrift, before finally being terminated in 1919.

The Audubonites mobilized once again, organizing editorial- and letter-writing campaigns in defense of the pelican nationwide. The government enlisted National Audubon Society secretary T. Gilbert Pearson for a fact-finding mission. He traced the Gulf's coastline—its 1,400 miles from the Mexican border to Key West—estimating a population of no more than 65,000 adult pelicans, not the hungry millions that fishermen augured. He poked around in pelican vomit—when frightened, the birds will often regurgitate their meal in order to lighten the load, so to speak, before flying away—and collected samples for lab identification. Pearson confirmed what any pelican devotee already knew: the species primarily subsists on a small, bony, superabundant swimmer unfit for human consumption—the menhaden, a forage fish whose seemingly inexhaustible supply has raised concerns of depletion only in the twenty-first century due to the increased popularity of fish oil. (The West Coast's pelican population, it should be noted, feeds on the once equally abundant, now equally overfished, anchovy and sardine.) Any drop in Gulf fish reserves, the Audubonites contended, stemmed from the excessive use of seine nets. Pelicans, and other sea surface–feeding fowl, actually help more than hinder "food fish" populations by driving menhaden and other forage fish to lower depths. The case settled, federal and state governments backed off from any anti-pelican legislation.

But the attacks continued. On March 2, 1924, members of the Halifax River Bird Club visited Pelican Island only to discover a massacre scene: over 1,200 dead juvenile pelicans, murdered sometime a month prior. Residents once again blamed the pelican for pilfering valuable food fish. One local chalked the crime up to the bird's "pure cussedness." The Florida Audubon Society posted a $100 reward, which, as far as I can find, still remains unclaimed. The following year, the pelicans abandoned Pelican Island.

* * *

Paul Kroegel died in 1948, having lived long enough to see his precious pelicans return home.

But the island has shrunk in size over the past century, losing half its acreage because of coastal erosion from storms and increasing boat traffic. Mangrove and shoreline restoration measures have abated further land loss, while making the island once again sustainable for birdlife. (Mangroves are also a leading mitigator of climate change. "On a per-hectare basis, mangroves are the ecosystem that sequesters the most carbon," according to marine biologist and conservationist Octavio Aburto. "They can bury around five times more carbon in the sediment than a tropical rain forest.")

In 1959, the state of Florida optioned the right to auction off the surrounding wetlands to property developers from Miami. Plans to dredge and fill the Indian River Lagoon horrified a local citrus grower named Joe Michael, who recruited fellow conservationists, lobbied state and federal authorities, and waged a four-year battle to save Pelican Island once again. Due to their efforts, the boundaries of the national wildlife reserve stretch far beyond the island to include nearly five-and-a-half thousand acres of federally protected land and water.

The reserve's highlight is Centennial Trail, a three-quarter-mile boardwalk comprised of planks engraved with the name, location, and founding year of each of the country's 568 national wildlife refuges, together encompassing over 150 million acres. Following the boardwalk, you travel backward in time—Kentucky's Green River NWR (2019) . . . New Hampshire's Umbagog NWR (1992) . . . Louisiana's Catahoula NWR (1958) . . . Puerto Rico's Culebra NWR (1909)—until you arrive at an observation tower overlooking the actual Pelican Island.

On my visit there in mid-December 2020, the island resembled a verdant, undulating mass of mangrove set adrift. This being our first year of COVID-19, I wouldn't touch the tower's pair of viewing scopes, but with my binoculars scanned the island and lagoon for birdlife. Gulls laughed and shrieked overhead. A pod of white pelicans buoyed on the salty, swampy sea. An osprey and great egret jointly perched on a mangrove—the strangest of marriages. No-see-ums tickled my arms. Fifteen minutes passed. And then, a single, unmistakably shaped head emerged from the island's very center. The brown pelican preened herself—three strokes left, three strokes right—fluttered her wings—once, twice, a third time to make sure I was watching—and lowered back down onto her nest, melting into the mangroves.

3

No Oases of Safety

In nature, nothing exists alone.

> —Rachel Carson, *Silent Spring*

The volcanic island of Anacapa rises rugged, windswept, and craggy just twelve miles off the Pacific Coast, due west across the Santa Barbara Channel from Los Angeles. The Indigenous Chumash people of central and southern coastal California named these three sea-cliffed islets *'anyapax,* meaning mirage or illusion, for the way the island appears to shift and change shape in the morning fog or afternoon swelter. Seabirds—auks, storm petrels, gulls—nest here by the many thousands, free from predation, surrounded by the ocean's bounty. Most importantly, Anacapa hosts the largest breeding colony of the California brown pelican, a subspecies of *Pelecanus occidentalis* differentiated from its Gulf Coast cousin by the brilliant reddish-orange

Opposite: Crushed DDT-exposed egg on Anacapa Island. Photograph by Betty Anne Schreiber Schenk.

hue of its pouch in breeding adults—so brilliant, in fact, that they've earned the nickname "red beards." If you've never thought of the species as a West Coast bird, consider the infamous prison island of Alcatraz, which takes its name from the archaic Spanish word for "pelican."

In the late 1960s, Robert Risebrough realized that the Anacapa pelican colony—one thousand pairs strong at last count—could hold the key to his research. A young, promising molecular biologist—as a Harvard grad student, he coauthored the 1961 messenger RNA article that helped James Watson win the Nobel Prize the following year—Risebrough was fascinated with genetic research and its literal minutiae. This interest in the microcosmic evaporated, though, after he discovered that the bald eagles he used to see swooping the Canadian shores of his Lake Erie boyhood had vanished. His thinking shifted from asking, "How could a biological structure—a human body or pine tree—come from a mixture of chemicals?" he told me by phone, to a realization that "that's less important than the fact that the pine forests and all the wildlife continue to live." Risebrough went macro, committing himself to conservation, and secured a position at the University of California's Institute of Marine Resources studying the effects of insecticides on seabirds.

On March 19, 1969, Risebrough landed on Anacapa with a small squad of ornithologists: Fred Sibley, Monte Kirven, and Joseph Jehl. The team located a nesting site of approximately 250 pelicans on a high ridge overlooking the north side of Anacapa's West Islet. Their goal, Risebrough said, was to "just grab a few eggs and come back down." After a strenuous climb to the top, they "suddenly halted in shocked dismay," Jehl recalled. Shattered egg fragments littered the ground like confetti. "It was immediately obvious," Risebrough wrote in his field notes, "that disaster had struck." Of the colony's 298 nests, each newly lined with fresh plant material, only twelve contained eggs, in clutches of ones and twos. Many nests cradled a single, sad, smashed egg. The men gathered five undamaged eggs and numerous fragments for analysis and headed back to their boat, appalled by what they had witnessed. When a follow-up expedition returned a few days later, not a single viable egg remained intact. "Every single egg was broken or crushed," Risebrough said. "Something was terribly wrong."

* * *

The first signs of trouble had appeared in the summer of 1956, not in California but in the Gulf of Mexico. Over three days in mid-June, Thomas Imhof, a notable Audubonite from Alabama, counted over twenty brown pelican corpses off the coast of Mississippi and Alabama. A few weeks later, a New Orleans couple birding along Isle au Pitre and the Chandeleurs tallied over fifty dead pelicans. The following year, concerned birders descended on North Island, Walter Anderson's reliable pelican hotspot in the Chandeleur chain, and were assured to find a healthy colony of several thousand birds.

Extending up from the toe of Louisiana's boot and constituting Louisiana's easternmost barrier islands, the Chandeleurs carve a northeasterly crescent of silt and sand deposited by the Mississippi River over the past two thousand years. The islands are a mishmash of semisolid islets, shifting sandspits, and what coastal geologists call "mud lumps," ranging in size from a few yards to several acres. Brown pelicans love nothing more than to loaf on these lumps. In 1918, ornithologist Alfred M. Bailey, then the curator of birds and mammals at the Louisiana State Museum, surveyed the Chandeleurs to determine if pelicans ate food fish or fish food. He expected to verify that pelicans ate menhaden, but did not anticipate how many pelicans he would find—50,000 birds, he estimated, the largest brown pelican colony in the world. "The ground was covered with eggs," he wrote, and "a veritable snowfield of swirling birds continually circl[ed] over head."

But in 1960, pelican counters revisited the Chandeleurs' North Island and discovered a greatly diminished population of just two hundred nesting pairs. Researchers returned two years later only to find a pitiful pod of six adult birds and not a single nest. In less than fifty years, the Louisiana brown pelican had become a pale reflection of its previous self.

Louisiana's participation in the Christmas Bird Count, an annual National Audubon Society–sponsored, volunteer-driven census started by Frank Chapman in 1900, likewise revealed a precipitous decline. Cameron Parish, by far the state's most diverse birding locale—it's aptly nicknamed "Louisiana's Outback"—submitted a count of ninety pelicans in 1956, nineteen the subsequent year, and not a single bird the year following. Rare sightings popped up here and there along the coast, but by 1963 the brown pelican had become what biologists call extirpated, or locally extinct, in Louisiana. The Pelican State had become pelican-less.

* * *

Whatever had decimated the state's pelican population was killing indiscriminately. Around the same time, anglers began reporting significant fish die-offs along the Mississippi and Atchafalaya river basins. Fishponds were emptied of all but floating fish. By 1963, an estimated five million fish were dying annually. "When we left Venice to run down the [Mississippi] river," Allen Ensminger, chief of the state's refuge system, recalled, "we'd be in dead fish the whole way." And it wasn't just fish. From Beaumont to Pensacola, great swatches of empty, birdless skies greeted visitors to the Gulf, in what scientists took to calling "blank spots." Towns throughout the country reported up to a 90 percent drop in songbird populations. Robins dropped dead after eating one meal's worth of earthworms. The bobwhite quail and wild turkey disappeared from Alabama. In Texas, sows birthed entire litters of dead piglets. Nursing calves died after their very first suckling. Other animal populations exhibited alarming mutations. DDT exposure caused gynandromorph, or the development of male and female characteristics, in mosquitoes. In Pacific Canada, coho salmon appeared to be going blind.

Olga Owens Huckins, among many other scientists and laypeople, had a hunch as to what was laying waste to so many different species. In late January 1958, the bird-loving Bay Stater wrote a letter to the *Boston Herald* detailing the aftermath of an anti-mosquito aerial spray campaign in her seaside town of Duxbury, Massachusetts. "The 'harmless' shower hath killed seven of our lovely songbirds," Huckins wrote. "All of these birds died horribly, and in the same way. Their bills were gaping open, and their splayed claws were drawn up to their breasts in agony." The culprit, Huckins surmised, was an insecticide named DDT.

Huckins penned a second letter to her friend Rachel Carson, a marine biologist turned celebrated environmental journalist. Carson had likely come across DDT before, back in August 1945, when U.S. Army-Navy scientists touted the pesticide as an "insect 'bomb'" that could do to mosquitoes and flies what nuclear weapons had done to Hiroshima and Nagasaki earlier that month. She dug in to research her next book.

First synthesized by Austrian chemist Othmar Zeidler in 1874, DDT—or dichloro-diphenyl-trichloroethane—remained an unheralded lab curiosity until Paul Hermann Müller discovered its pesticidal applications sixty-five years later, a breakthrough that would win the Swiss chemist a Nobel Prize. DDT is classified as a chlorinated hydrocarbon, or, in organic chemistry terms, a compound containing chlorine, hydrogen, and carbon. If insects wrote chemistry textbooks, they'd say DDT is a synthetic poison that causes their neurons to spontaneously fire, leading to spasms and eventual death. During World War II, Allied forces dusted troops, civilians, and prisoners with DDT powder to combat typhus-carrying lice in Europe. The Army blanketed the South Pacific Islands with the insecticide in hopes of eradicating the mosquito-borne malaria and dengue fever. Their efforts came to fruition when scientists determined that, though insoluble, DDT could be mixed with oil or gas and sprayed on top of water, where it would destroy mosquito eggs before they hatched. By war's end, numerous insecticide manufacturers marketed DDT for the home and garden; fog trucks traced suburban streets, farmers crop-dusted their fields. In the early 1950s, the nation annually produced over 100 million pounds of the chemical, despite numerous worries and warnings about its toxicity to plants, animals, and humans. It zapped flies, mowed down moths, and put bed bugs permanently to sleep, but, as one homemaker's magazine warned, "DDT presumably could send *you* on a death jag too."

Few books have altered the course of human history quite like the exposé that culminated from Rachel Carson's research into the uses and abuses of DDT. "This is a book about man's war against nature," she wrote, "and because man is part of nature it is also inevitably a book about man's war against himself." Serialized in the *New Yorker* and *Audubon Magazine* in the months leading up to its September 1962 publication, *Silent Spring* condemned not just DDT, but over two hundred synthetic pesticides and herbicides—"agents of death," in Carson's words—formulated since the mid-1940s. Endrin ranked as the most toxic, three hundred times more poisonous to birds than DDT. Used to liquidate corn borers in the Midwest, cotton bollworms in the Upper South, and sugarcane borers throughout Louisiana, endrin (alongside the fire-ant exterminators heptachlor and dieldrin) washed down the Mississippi River, resulting in all of those fish kills. "It is not possible to add pesticides to water anywhere," Carson warned, "without threatening the purity of water everywhere."

Big Ag and Big Chem quickly mobilized to discredit and destroy Carson, attacking her authority, her character, and, most predictably, her gender. Former secretary of agriculture Ezra Taft Benson sent a letter to Dwight Eisenhower, maliciously asking "why a spinster with no children was so concerned about genetics," before answering his own question: she was "probably a Communist." *Silent Spring,* the National Agricultural Chemical Association alleged, was "more poisonous than the pesticides she condemned."

Carson knew how to make the science of chlorinated hydrocarbons a social issue. We should think of DDT, endrin, and the like not as pesticides, she argued, but as biocides—annihilators of all life—for the way they work up the food chain in a process called "biological magnification." These biocides leach into waterways, where they are incorporated into the diets of zooplankton, which are gobbled up by fish and invertebrates, which then may be eaten by seabirds, raptors, and other predators, like humans. As these poisons jump from species to species they biomagnify, or accumulate in progressively greater concentrations, especially within tissues and organs rich in fatty substances, where they can become carcinogenic. One of *Silent Spring*'s most alarming excerpts outlined a study that found increasing traces of DDT residue in human breast milk, able to be passed on from mother to child. Our world has become a "poisoned environment," Carson wrote, and "no oases of safety remain." By extension, the same could be said of ourselves.

The age of the Anthropocene—a geological epoch defined by global climate change, widespread pollution, the mass extinction of species, and other horrors wrought by humanity's impact on the earth—had fully arrived.

* * *

Down in the now pelican-less state of Louisiana, scientists and ornithologists knew that Carson's anti-pesticide polemic could explain a decade of fish and bird die-offs in the state. Two decades' worth of tainted Mississippi River drainage-basin runoff had contaminated the Gulf of Mexico. In October 1967, Louisiana Wildlife and Fisheries Commission chief Leslie Glasgow wrote to Alexander "Sandy" Sprunt, the National Audubon Society's research director who had been studying the links between the decimation of the nation's bald eagle population and DDT. "We are interested in exploring the possibility of restoring our state bird, the brown pelican, as a breeding species," Glasgow communicated. "So far we have done nothing except talk." Sprunt pleaded the pelican's case up the bureaucratic chain to John Gottschalk, director of the U.S. Bureau of Sport Fisheries and Wildlife. *If you want to help the pelican,* Gottschalk pushed back, "Why not do something about it?"

Glasgow and Sprunt resolved to transform chatter into action, calling a three-day summit to be held in mid-January 1968 at Louisiana's Rockefeller Wildlife Refuge in Grand Chenier. There, twenty-four federal, interstate, and private wildlife authorities that called themselves the "Pelican Committee" hatched one of the wildest plans in the annals of wildlife conservation. The Louisiana Wildlife and Fisheries Commission—known today as the Louisiana Department of Wildlife and Fisheries—would be in charge of capturing semi-fledged pelicans, aged nine to twelve weeks old (think of them as awkward, just learning to fly, teenagers), from a healthy breeding colony and relocating them to the Louisiana coast. This moonshot proposal, the translocation of a locally extinct species of bird, hadn't ever been successfully achieved. Though brown pelicans sometimes exhibit erratic migration patterns—they will fly as far as necessary to find food—they are a largely philopatric, or permanent residence–seeking, species. The committee could count that as a plus. But pelicans, like so many birds, learn their singularly birdish skills through a combination of instinct and imitation. Could teenage pelicans reliably learn to dive for fish on their own?

The brown pelican translocation project would be headed by two biologists who, on paper, had no business messing with this bird. Ted Joanen and Larry McNease were not ornithologists but crocodilian conservationists working to rehabilitate the state's waning alligator population through captive breeding efforts. But the pair—lifelong Louisianians both—knew the state's coastal waters and wildlands better than most anybody. If they could successfully save Louisiana's alligator from extinction, maybe they could bring the state's brown pelican population back from the dead.

Joanen and McNease began scouring the Gulf Coast for purloinable pelicans. Mississippi and Alabama had never possessed notable pelican breeding grounds and Texas's population, like Louisiana's, had been recently wiped out, so they looked to Florida's Atlantic Coast, where the lack of any significant watershed outlets limited the spread of pesticidal contaminants. First they secured the help of Ralph Schreiber, a brilliant young ornithologist doing graduate field-work in Tampa and on his way to becoming the world's leading expert on brown pelicans. They then trucked down to Cocoa Beach—not far up the Indian River Lagoon from Pelican Island National Wildlife Reserve—bagged fifty birds, and returned to their home base at the Rockefeller Wildlife Refuge: a sprawling, state-run habitat skirting the southwestern Louisiana coast.

After a two-month quarantine, the team released twenty-five banded birds 160 miles east of the refuge on Grand Terre Island, an idyllic, dune-humped, mangrove-rich islet separated by a narrow channel from the fishing community of Grand Isle. They clipped the wings of ten members of the group to encourage the remaining fifteen to not take flight to Florida. The plan worked. The free-flyers scoured the surrounding waters of Barataria Bay for nourishment, while the flightless gorged on a free fish buffet provided by their captors. The committee freed the remaining twenty-one pelicans—fifteen flyers, six flightless (four did not survive the odyssey)—at the Rockefeller Refuge in October. Once again, the young birds successfully adapted to their new surroundings, until a series of severe storms wrecked the colony the following March. Within weeks, all twenty-one Rockefeller pelicans had perished. A state lab analysis found significant levels of insecticide residues in the dead birds: dieldrin, endrin, DDT, and its breakdown products, DDE and DDD. The combined traumas of tempest and toxicants had leveled the birds.

Joanen and McNease didn't give up. The following year, 1969, they abducted another fifty-five semi-fledglings from Merritt Island and followed the established pattern, splitting the pack between Grand Terre and Rockefeller. The same results came back: the Grand Terre population thrived; the Rockefellers couldn't survive winter's end. The group doubled down in 1970, taking one hundred birds and releasing them all on Grand Terre. The team members were experts by now in the art of birdnapping. They discovered that young pelicans have the nerves of a prom-night wallflower and will regurgitate any fish hand-fed to them, no matter how famished they may be. Joanen developed a recipe for pelican pablum, a slurry of canned cat food and water, that went down easy. The team also learned, through plenty of trial and error, the exact right time to snag a pelican: just days before they are fully fledged and flying and feeding themselves. "There's a big difference in the strength of a pelican before he flies and eats and *after*," Joanen recalled in amusement. "We returned to Louisiana looking like we'd been fighting with a roll of barbed wire."

After three years, the committee had determined how to build a pelican colony from scratch. But for all the team's hard work, that Grand Terre colony of 150 birds appeared uninterested in the business of making pelican babies. The disparate breeding seasons of Atlantic Seaboard pelicans (autumn/early winter) and those from the Gulf Coast (late winter/early spring, to avoid the hurricane season) raised the question of whether the transports would even make an attempt at mating. The biologists knew that the following season would mark the pod's sexual maturity midpoint of three to five years. Unless the birds nested, Louisiana could not count on having a viable pelican population anytime soon.

* * *

Back on the West Coast, Robert Risebrough and his ornithological colleagues rushed to save the California brown pelican from the same fate as Louisiana's state bird. Those crushed eggshells, they knew, held the key to understanding the pelican's nationwide decline over the past decade. Just five months before Risebrough's Anacapa research trip, *Science* magazine published a paper linking the plummeting populations of raptors (taloned birds of prey like falcons, eagles, and osprey) *and* carnivorous, non-raptor seabirds like gulls, cormorants, and pelicans. Coauthored by Joseph J. Hickey— the hand-selected heir of America's conservationist godfather, Aldo

Leopold—and Daniel Anderson—his University of Wisconsin graduate student researching ecotoxicology in waterfowl—the article argued that catastrophic declines of several species had been accompanied by a measurable decrease in eggshell weight and thickness. The numbers were significant and shocking. The bald eagle's eggshell weight had dropped by 20 percent since the 1940s. Over the same period, the average thickness of osprey and peregrine falcon eggs thinned by over a quarter. Herring gull nests in Wisconsin contained flaking eggs a full third smaller than in previous years.

Though Anderson wanted to stay in the Midwest to study ducks, Hickey had other plans for his advisee, sending him on a year-long research trip to measure eggshells. "I spent a lot of time in stuffy old museums," Anderson told me in a phone conversation. *A lot of time* is a gross understatement. He visited every major natural history museum in North America, plus many in Europe, and even the collections of private oologists who traded in black-market eggs. "I became good friends with some pretty illegal guys," he laughed, "egg collectors, bad boys." Armed with a scale and micrometer, Anderson surveyed upwards of 50,000 eggs, finding the same general trend of shell thinning, beginning, unsurprisingly, in 1947 with the proliferation of chlorinated hydrocarbon insecticides. "It goes back to Rachel Carson," he told me. "I used to tell my students, we're all the grandchildren of Rachel Carson."

Anderson's data sets jolted the pelicanologists. Anacapa eggshells measured at half the thickness of eggs collected from the same island before World War II. (Compare that to egg thinning numbers out of Florida, where the thickness of pelican shells dropped by just under 10 percent during the same period.) Further lab results showed that the soft tissue of Anacapa eggs contained sixty-eight parts per million (ppm) of the DDT byproduct DDE, or twenty-seven times more than the critical level of 2.5 ppm. Because pelican parents jointly incubate by essentially standing atop the nest clutches with their webbed feet, their eggs stood no chance of survival.

Risebrough didn't have to look far to locate the source of all that DDT. The Montrose Chemical Corporation, the nation's largest manufacturer of the insecticide, inhabited an inconspicuous industrial corridor in the beachfront city of Torrance, south of Los Angeles. Beginning with the factory's opening in 1947, Montrose discharged DDT into the Los Angeles County sewage system, which emptied directly into the ocean. Additionally, the company jettisoned 2,000 barrels of DDT-laced acid sludge every month for nearly fifteen years at a dumpsite ten nautical miles northwest of Catalina, on the way to Anacapa, and in shallower waters, just off the coast, which today ranks as the nation's largest underwater Superfund site. Combined, the chemical company dumped more than an estimated 2,500 tons of DDT toxic waste into Pacific waters. Half a million barrels of that deadly ooze still reside on the seafloor, and have slowly been seeping their contaminants into the ocean for over the past half-century. As I write this, a recently released report blames these same "legacy chemicals" for a massive cancer outbreak in southern California's sea lion population.

By the beginning of the 1970s, due substantially to the work of Risebrough, Anderson, and other ornithologists, the USDA had canceled the application of DDT for most large-scale and household uses. A lawsuit brought by the Environmental Defense Fund, an organization originally founded to safeguard birds from DDT, forced the federal government to finally outlaw the pesticide in 1972. But many feared that the ban arrived too late to save the brown pelican. In 1969, the U.S. Endangered Species Conservation Act had declared the bird to be "threatened with extinction."

* * *

The winter of 1971 brought welcome news to the Louisiana Pelican Committee. Late that year, a handful of Grand Terre transplants built thirteen nests on nearby Camp Island, a shallow, shell reef cay susceptible to high tides. Two previous nesting attempts had been washed out to sea. But brown pelicans are nothing if not tenacious when it comes to procreation. Blessed with calm currents, their third try would prove to be a success, and within a month eight baby brown pelicans appeared, the first Louisiana-born birds in a decade.

The following year produced fourteen fledglings, a promising but paltry number considering storms had swept several dozen eggs into the Gulf. As 1973's breeding season approached, scientists hung aluminum pie tins from wires strung across Camp Island, with the hopes of making its shores annoyingly un-nestable. (The island has since been completely submerged by storm-borne waters.) The pod resettled a bit farther into Barataria Bay, on a long-established wildlife preserve and oyster reef called Queen Bess Island, and fledged twenty-six pelicans. The next season, the Queen Bess colony produced 104 new birds. Louisiana was once again in the business of breeding brown pelicans.

But a disastrous die-off struck Barataria Bay in 1975, plummeting the total population by 40 percent and sending the committee into anxious fits. Autopsy reports revealed pelican brain tissue and eggs riddled with traces of eight different pesticides, including DDT and endrin. Despite the wholesale banning of the former and the dwindling use of the latter, poisonous organochloride residues may be found in waters worldwide up through the present day.

Notwithstanding this setback, the Queen Bess colony continued to flourish, producing so many birds—over 6,000 in total by 1990—that Joanen and McNease started transplanting juvenile pelicans to Raccoon Island, the westernmost outpost in Terrebonne Parish's Isle Dernière archipelago. The transplantation, beginning in 1984, continued for three years. Over the next two decades, those 149 birds bred over 100,000 more brown pelicans, making Raccoon the state's largest and most productive colony.

The Pelican Committee made one more, perhaps largely symbolic, attempt to rehabilitate Louisiana's state bird. Over four years in the late 1970s, they birdnapped several hundred nestlings from Florida's Gulf Coast and released them on Isle au Pitre and Walter Anderson's beloved North Island, formerly the world's largest brown pelican colony. Their efforts succeeded, and platoons of pelicans soon staked claim to the Chandeleurs' mud lumps to fish for food. By the next decade, especially after the banning of endrin in 1984 and dieldrin three years later, the brown pelican once again patterned the sky over all five Gulf Coast states.

On November 9, 2009, forty years after the first of 1,276 total Florida fledglings landed on Louisiana's shores, and with the California population steadily recuperating due to the banning of DDT (Anacapa

Island averages around 5,000 nesting attempts annually), the federal government removed the brown pelican from the endangered species list. "At a time when so many species of wildlife are threatened, we once in a while have an opportunity to celebrate an amazing success story," Secretary of the Interior Ken Salazar proclaimed. "Today is such a day. The brown pelican is back!"

* * *

But with the pelican's revival came retaliation. Just as they had in the early twentieth century, commercial fishermen blamed the bird for any downturns in yield and persecuted the brown pelican in far more savage ways than before.

In late 1982, at Dana Point, south of Los Angeles, twenty pelicans washed ashore with their top mandibles hacked off. The following year, at nearby Redondo Beach, a pair of bait fishermen electrocuted over a dozen pelicans for stealing their anchovies. In Huntington Beach the same year, the courts levied a $1,200 fine on a boat captain who tortured and killed a pelican with a pair of pliers. Farther north in Monterey, a serial slasher mutilated the beaks of at least thirty pelicans.

Scientists soon blamed El Niño, the climatic condition that occurs when bands of warm ocean water develop off the equatorial Pacific coast of Latin and South America before tracking north, causing—among many other effects—pelican provisions like herring and anchovies to flee the California coast.

The Gulf Coast is not immune to such atrocities. Across the eastern Sunbelt, the brown pelican is simultaneously embraced as a symbol of countless coastal communities and a pariah, a nuisance, a scapegoat

Above: Hatchling pelican on Raccoon Island. *Opposite:* Nesting brown pelican and chicks on Raccoon Island. Photographs by Liz Bourgeois.

for seas suffering from overfishing and man-made climate shifts. Every few years, especially in Florida, where pelicans live in intimate proximity to large cities, a rash of pelican mutilations and murders occurs.

A decade following the first wave of attacks on pelicans, a fresh surge of violence heralded another severe El Niño cycle. "Words can't describe what was going on," one seabird rehabilitator told the *Los Angeles Times*. "We were constantly sewing up pelicans, changing bandages on pelicans, feeding pelicans. It was like working in a MASH unit." Men killed pelicans with their bare hands. Along the coast, north and south of Los Angeles, animal rescue volunteers found numerous pelicans with their beaks sawed off. In Newport Beach, police discovered a young pelican pinned to a light pole, crucified alive.

Above: California brown pelican and Louisiana brown pelicans on Raccoon Island. Photograph by Liz Bourgeois.

Opposite: Nestling pelicans on Philo Brice Island. Photograph by Rien Fertel.

4

Insatiable Creatures

No other Birds so grand we see!
None but we have feet like fins!
With lovely leathery throats and chins!
Ploffskin, Pluffskin, Pelican jee!
We think no Birds so happy as we!

—Edward Lear, *The Pelican Chorus and Other Nonsense*

On April 20, 2010, five months after the brown pelican's removal from the nation's endangered species list and roughly fifty miles southeast of the Mississippi River Delta, the Deepwater Horizon drilling rig exploded, sending a fireball roaring into the Gulf of Mexico's once quiet twilight. For the next eighty-seven days, the BP-operated Macondo Prospect well leaked approximately 134 million gallons of crude oil into the Gulf's waters—the largest marine petroleum spill in human

Opposite: Oiled brown pelican rescued from the Deepwater Horizon oil spill. Photograph by Kim Betton, U.S. Fish and Wildlife Service.

history. The Deepwater disaster occurred at the very worst time for local birdlife, in the thick of the breeding season. Scientists have concluded that the crude, in addition to the nearly two million gallons of toxic dispersant chemicals like Corexit sprayed on the water's surface, killed countless fish, sea turtles, marine invertebrates, and mammals, as well as an estimated one million coastal and offshore birds (though one survey puts that number as high as 2.5 million), including around 10 percent of the northern Gulf's brown pelican population.

Footage from those dark days of summer is likely forever ingrained in your memory. Seabirds slick and slimy with oil, their pink and white feathers stained a chocolate brown. Dead birds belly-up in sludge. Brown pelicans struggling to lift their wings, which drip heavy with muck the color of clotted blood. Exposure to petroleum damages the delicate plumage layers that allow for buoyancy, insulation, and flight, causing birds to die from drowning, hypothermia, starvation, and dehydration. Then there are the long-term chemical effects, known and unknown, that scientists will be studying in the coming decades: cancers, metabolic disorders, reproductive diseases.

It's worth noting that oil spills have devastated pelican populations since tankers first plied the seas. In 1921, off Florida's east coast, petroleum "spread a deathtrap over the waters in which thousands of birds [met] their fate," according to an editorial in the National Audubon Society's newsletter. "A Brown Pelican, that looked as though it had been dipped in a tar-barrel, was subject for the kodaks [*sic*] of thoughtless tourists at Daytona Beach who seemed not to realize the bird's hopeless plight."

The pelicans and people who call the Gulf Coast home will be living with the BP oil spill's aftershocks for generations, but one bird in particular, documented in a widely disseminated photograph, haunts me to this day. Captured by Kansas City–based AP photographer Charlie Riedel on June 3, it shows a crude-covered pelican standing in undeniable agony on the shores of Grand Terre Island, the original site of the pelican translocation project. The image bears an uncanny resemblance to the notorious "Hooded Man" portrait that surfaced out of Abu Ghraib prison in 2003. Both the pelican and the hooded Iraqi detainee are shrouded—the former in oil, the latter in what appears to be a thin rug—rendering their bodies amorphous, unnatural. In these twin scenes of torture, bird and man rise from the earth, heads cocked in pain, wings and arms outstretched in an unmistakably Christlike pose, welcoming the world to the horrors of a brand-new century, and daring it to look away.

* * *

For almost as long as humankind has attached anthropomorphic meaning to animals, we have portrayed the pelican as bridging the heavenly realm and this earthly life of suffering. Ancient Egyptians associated the pelican with death and the afterlife and called upon the bird as an usher into the great unknown. The Papyrus of Nu, from the *Book of the Dead,* implores a pelican goddess named Henet to invite the deceased into the infinite bounds of her tomblike pouch: "The doorwings of the heavens stand open for me. / The doorwings of the earth stand open for me. / The mouth of the pelican is opened for me."

In contrast, the Torah describes the pelican as a lonely, wandering, worldly creature classified, alongside carrion vultures, as an unclean abomination. The Hebrew word for pelican is *qaath,* or *kaath,* meaning "to vomit," an apt name for a bird that feeds its chicks by regurgitating partially digested fish into the nest.

Around the same time as the Torah's composition, the Greek fables accredited to Aesop tell of a meeting between the ostrich and the pelican, who sports a blood-soaked breast. "What accident has befallen you?" the ostrich asks. "You certainly have been seized by some savage beast of prey, and have with difficulty escaped from his merciless claws."

No, the pelican laughs, "I have only been engaged in my ordinary employment of tending my nest, of feeding my dear little ones, and nourishing them with the vital blood from my bosom."

"Is this your practice," the ostrich counters, "to tear your own flesh, to spill your own blood, and to sacrifice yourself in this cruel manner to the importunate cravings of your young ones?" Do like me, she advises, lay your eggs upon the ground and lightly cover them with sand. "If they have the good luck to escape being crushed by the tread of man or beast, the warmth of the sun broods upon, and hatches them, and in due time my young ones come forth. . . . I leave them to be nursed by nature."

"Unhappy wretch," says the pelican, "who knows not the sweets of a parent's anxiety; the tender delights of a mother's sufferings! It is not I, but thou, that art cruel to thy own flesh."

Early Christian texts expanded on Aesop's fable, transforming the once profane pelican into something sacred, an association that dominates perception of the bird to this day. The earliest notable reference comes from the *Physiologus*—from the Greek word for "naturalist"—an anonymously authored, second-century compilation of allegorical animal tales that, once translated and disseminated, became one of history's first popular books. Pelican babies, the *Physiologus* explains, once grown but not yet ready to leave their nest, tend to strike their pelican parents in the face with their beaks. Bewildered and more than a little disgruntled, the adults hit back, killing their young ones. Filled with grief and moved by compassion, the mother pelican weeps over her slain brood for three days. On the third day, with wings spread wide, she sword-plunges her beak into her side, piercing flesh to spill blood upon her babies, who are soon revived. A gruesome illustration accompanies the story: torrents of blood spurt forth from the mother pelican's breast, dousing her chicks with gore. "The pelican," the *Physiologus* explains, "is an exceeding lover of its young."

For those who paid attention during Bible study, the only thing missing from this Eucharistic metaphor is a Communion wafer. But for any medieval-era heathens unable to delineate the Christly comparisons, the *Physiologus* clarifies: "The Maker of every creature brought us forth and we struck him. How did we strike him? Because we served the creature rather than the creator." The cross-like wings; sword-like beak, side wound, and blood; death, sacrifice, and resurrection; love and suffering—the parable of the pious pelican is complete.

Medieval bestiaries, or illuminated compendiums of real and imagined beasts, expanded on the *Physiologus*'s basic outline. Frequently it's a crow, or more often a snake—representing Satan, naturally—that slays the poor pelican nestlings, as in *De proprietatibus rerum* (*On the Properties of Things*), a thirteenth-century encyclopedia penned by the Franciscan monk Bartholomaeus Anglicus. "The serpent hateth kindly this bird," he writes. "Wherefore when the mother passeth out of the nest to get meat, the serpent climbeth on the tree, and stingeth and infecteth the birds. And when the mother cometh again, she maketh sorrow three days for her birds."

Some stories claim that the father kills his young, which the sorrowful mother resurrects; in other tellings the roles reverse, and only paternal care can revive the brood. Often these holy fables speak of pelican chicks who die from too much love, smothered by a mother's affection. Guillaume le Clerc's *Bestiaire divin* (*Divine Bestiary*) even aspired to raise the pelican's status higher than that of God's lowly but beloved ewe. "Never did mother-sheep love her lamb," he insists, "as the pelican loves its young."

Like the *Physiologus,* these texts contain grisly illustrations of blood-splattered pelicans, each more Tarantinoesque than the last. The Catholic Church refers to this image as the "vulning pelican"— from the Latin *vulnerō,* "to wound"—and the "Pelican in her Piety." Gentler, less sanguinary depictions—often with the crown of thorns symbolically substituted for the pelican's nest—fill chapels, churches, and cathedrals across Christendom with devotional carvings, mosaics, paintings, stained-glass windows, and gargoyles (Notre Dame Cathedral has a superb example of the latter). In *Paradiso,* Dante speaks of Christ as "*il nostro Pelicano.*" In his venerated Eucharistic hymn "Adoro te devote" (1264), St. Thomas Aquinas summons his "Good Lord Bird" with the following: "Lord Jesus, Good Pelican, / clean me, the unclean, with Your Blood, / One drop of which can heal / the entire world of all its sins."

The suffering pelican symbol was not reserved for Christ alone. *The Aberdeen Bestiary,* dating to the twelfth century, classifies the most pious of Christly sufferers as worthy of the pelican honorific. Just as the bird sacrifices her own body for the greater good, the manuscript counsels, "when we concern ourselves less with matters of flesh and blood and concentrate on spiritual acts, by conducting ourselves virtuously," we can become the pelican.

Pursuing that same theme, children's stories often portray pelicans carrying or cradling the young in their pouches, like Nigel, the brown pelican costar of *Finding Nemo*—an update of the stork-as-baby-deliverer trope. These pelicans are not only motherly, but often act as a child's surrogate guardian, and thus Christlike savior, as in literature's most famous pelican story: Colin Thiele's *Storm Boy* (1964). In this tearjerker of a tale, the eponymous Storm Boy rescues a trio of orphaned pelican nestlings on a desolate Australian seacoast. Storm Boy releases the siblings after nursing and fledging them, but his favorite, Mr. Percival, remains behind. The pelican becomes the motherless boy's best friend, substitute parent, and a local hero, who rescues six shipwrecked sailors and later sacrifices his own life to safeguard a flock of ducks from hunters. One of the book's lessons succinctly summarizes our relationship with the pelican. "In the world there will always be men who are cruel," Storm Boy's father tells him, "just as there will always be men who are lazy or stupid or wise or kind."

Like many myths, the "Pelican in Her Piety" remixes untold millennia of observational misunderstandings into an easily digestible narrative. Theories abound. For instance, the pouch of the Dalmatian pelican, whose wide range stretches over much of the central Eurasian landmass where the vulning legend propagated, turns a bright blood-red during its breeding season. Likewise, the breeding breast plumage of Africa's great white pelican turns a rose-pink tinge. Several pelican species have a red-orange bill tip, likely an evolutionary bullseye of a visual cue—also common to some gulls—for chicks to peck and poke at when hungry. When a pelican rubs that red pouch or bill tip across its breast, whether while preening or self-Heimliching to disgorge that last bit of regurgitated food for its nestlings, the display could be mistaken for a ritual bloodletting.

By the Renaissance era, Shakespeare and other writers had fully claimed the pelican as a secular symbol denoting charity and self-sacrifice, often to the point of self-detriment. The bird appears in three of the Bard's plays, including *King Lear,* whose titular character refers to his two oldest, ungrateful children as "pelican daughters"—it's understood that they greedily bleed him dry—and *Hamlet,* in which Laertes swears revenge in the bloodiest of terms: "To his good friends thus wide I'll ope my arms / And, like the kind life-rend'ring pelican, / Repast them with my blood." The Romantic poets took this subjective secularism one step further, transforming the pelican into a masochistic symbol representing the suffering artist, the bleeding poet. Goethe likened the act of writing to "a creation which I, like the pelican, fed with the blood of my own heart." The French playwright and poet Alfred de Musset compared the artist's tortured existence to the pelican's "sacrifice divine," spilling blood "in a love sublime," while Lord Byron challenged his haters with a rhetorical retort: "Would you know why I have done all this? / Ask of the bleeding pelican why she / Hath ripped her bosom? Had the bird a voice, / She'd tell thee 't was for *all* her little ones."

*   *   *

Just over six months into the Deepwater Horizon crisis, with first responders still patrolling the Gulf tallying dead birds and netting oiled pelicans for rehabilitation, the state of Louisiana unveiled its new flag and seal. For those Louisianians not paying close attention, the redesigned state symbols looked very much the same: a traditional Pelican in Her Piety and her chicks floating above the state motto, "Union, Justice, Confidence."

Few state—or, dare I say, national—flags compare to the one that flies over Louisiana. My opinion is biased, but you don't have to be a follower of Christ to appreciate the sorrowful beauty of the vulning mother pelican backdropped in blue. Granted, the design disregards two, arguably three, of the five basic principles of flag design as outlined by the North American Vexillological Association:

Keep it simple: Nope, pelicans are rarely simple, and neither is the pelican flag.

Use meaningful symbolism: For the win!

Use two to three basic colors: White and blue with a golden-brown dab—check!

No lettering or seals: Big fail on this one.

Be distinctive but use similarities to show connections: Distinctive, yes, but lacks the stars, stripes, etc. that most state flags incorporate. Though, as we shall see, an eventual touch of crimson will embody the good old red, white, and blue.

For the record, a 2001 NAVA survey ranked Louisiana's pennant 36th out of seventy-two flags belonging to states, provinces, and territories of the United States and Canada (New Mexico's simple and striking red and gold Indigenous sun won out).

Louisiana didn't always hold the brown pelican in such lofty regard. After the Louisiana Purchase, Governor William C. C. Claiborne chose the bald eagle to grace the state seal, an indication that the territory's allegiance lay with its new, American master rather than its French and Spanish colonial forebears. After Louisiana became the nation's eighteenth state in 1812, Claiborne replaced the eagle with a pelican, likely to appease the mostly Catholic citizenry. Elsewhere, Americans appeared nonplussed. "The people of the new State have strange ideas," a Nashville newspaper grumbled soon after. The new seal portrayed a Pelican in Her Piety looking more than a little vulture-like and overburdened with a litter of ten hungry chicks. Except for the brief Civil War interregnum, when reactionary rebels adopted a bland yellow star and thirteen-striped eyesore, at least a half dozen different iterations of the pious pelican have decorated the state's seal and flag. Bizarrely, the state legislature never got around to officially codifying the design. Sometimes the pelican faced left, sometimes right. Nestlings numbered anywhere between three and eighteen. In one flag, the pelican hovered over her nest, appearing totally unprepared for the demands of motherhood. Sometimes her beak was knife-sharp, other times blunt and bulbous. One common design portrayed a cute and cuddly pelican, ready for a starring role in the next animated Disney film. In another, her wings looked to be made of melting wax, like some Gulf Coast Icarus. Most confusing of all, each and every one of these birds appeared to be an American white pelican, a species that winters in Louisiana but does not call the state home year-round like the brown pelican.

After dragging their feet for a century and a half, legislators made the pelican—curiously, a generic, non-species-specific pelican—the state bird in 1958. Seven years later, a revised amendment adopted the brown pelican as the official state bird. Ironically, it would take another decade for brown pelicans to once again take up residence in the state.

But Louisiana's flag problems were not finished. In 2005, David Joseph Louviere, a teenager from Houma, Louisiana, noticed a continuing discrepancy in various designs. His flag, one that had formerly flown over the state capitol, featured three ruby droplets of blood on the mother's pierced breast. He noticed that most other flags flying across the state, including those flapping around the capitol parade grounds, were entirely bloodless, a sneaky cost-saving measure. He presented his findings in a social studies class project, then wrote a letter to his state representative in late October, just two months after Hurricane Katrina and the Federal Floods devastated the state. Louisianians deserve to be united under a standardized flag, Louviere testified in Baton Rouge the following year, "especially during a time like this when Louisiana is in recovery."

Lawmakers swiftly passed a resolution calling for "an appropriate display of three drops of blood." Those tasked with redesigning the seal and flag settled on a tidy, triangular pattern covering the bird's right breast. They also added a golden-brown tuft of feathered flame atop the mother's crown and a similar wash of color on all four birds' lower mandibles to make the family favor the brown pelican rather than the American white. Arguably, she *still* resembles the latter— "Our brown pelican is white," wrote one of the state's leading newspaper columnists—but at least her piety had been restored.

* * *

We've come this far—sixty-some-odd pelican-filled pages—without a single reference to the most well-known (and well-worn) slice of pelicana. So, without further ado—and with sincere apologies to all poetry lovers and pelicans—I present Dixon Lanier Merritt's limerical ode to the bird:

A wonderful bird is the pelican,
His bill will hold more than his belican,
He can take in his beak
Food enough for a week,
But I'm damned if I see how the helican!

I can't count how many people who, upon hearing that I was writing a book about pelicans, broke into Merritt's poem like Lin-Manuel Miranda dropping his first lines in *Hamilton.* I much prefer the short, sweet, and no less poetical verse composed by wildlife artist Charley

Harper: "If your food is all finned and your chin's double-chinned, you're a Brown Pelican. The seine with a brain."

That seine—an elastic, three-layered bag of skin, muscle, and mucous membranes called the gular pouch or sac—has confounded humanity since it first spotted a pink-backed pelican winging its way across the shallow swamps of sub-Saharan Africa. *What goes on in there?* we ask. *Enough food for an entire week? In his beak? Is it an extra stomach? And why does it look like—well, let's be honest—a certain component of the mammalian male anatomy? Does he fly with that thing all swollen and saggy with water?*

Aristotle suspected that the pelican's inner workings contain a built-in furnace capable of shucking meat from shell. "Pelicans that live beside rivers swallow the large smooth mussel-shells," the Greek philosopher surmised in *The History of Animals.* "After cooking them inside the crop that precedes the stomach, they spit them out, so that, now when their shells are open, they may pick the flesh out and eat it." Pliny the Elder believed that pelicans have a second stomach in their throats, which the "insatiable creatures" fill before passing the undigested nourishment to the belly's "true stomach." Pelicans do not eat shellfish, nor are they endowed with such super stomachs, though internet chatter persistently and falsely claims that pelicans have three bellies, the last dedicated to the digestion of fish bones. (All pelicans, let's note, have two stomachs: an enzyme-producing glandular organ called the proventriculus and a fish bone–grinding gizzard—an anatomy combo no different than that of your run-of-the-farm chicken.)

Confronted with such an odd eater, man has frequently (and cruelly) tested the bird's gustatory limits. The nature writer George S. Fichter tells of tossing baitfish after baitfish into a Florida pelican's maw, until "literally stuffed to the gills, he fell over when he tried to walk." After a fifteen-minute nap, the Falstaffian fellow, "as hungry as ever," began begging for more food. This story reveals more about the brown pelican's unfortunate status within fishing communities as a semidomesticated dockside moocher—take this as a gentle nudge to not hand-feed pelicans, dear reader—than it does about the bird's supersonic metabolism. Though it is worth noting the pelican's immense caloric needs: up to four pounds of food daily, half its average adult body weight (between six and seven pounds, according to one population survey). Compare that to the bald eagle, an eleven-pound bird that requires no more than a pound of food per day.

Unlike its sister species, who scoop up sustenance while floating along the water's surface, the much smaller and lighter brown pelican is the only member of the Pelecanus family that plunge dives for its prey. To watch a pelican fish for food, plummeting with seemingly kamikaze boldness from perilous heights, is to glimpse one of the avian kingdom's grand spectacles.

Whether solo foraging or hunting in packs, a pelican will scan the water below, flying slowly, usually upwind, constantly correcting her speed, in search of an individual fish rather than an entire school of swimmers. She nosedives beak first toward her intended quarry, thirty to sixty feet up on average (but sometimes even higher), wings expanded and feet forward to allow for adjustments, all the while subtly spiraling to the left in order to protect the esophagus and trachea located on the right side of her neck. Striking the surface at a steep sixty- to ninety-degree angle at up to forty miles per hour, wings and feet retract. Subcutaneous air sacs along the brown pelican's throat, breast, and underwings act like bubble wrap, cushioning the dive's impact while providing buoyancy. (Tall tales circulate of pelicans rendered blind from striking the water. This is a myth, though polluted seas, from chemical spills and the like, can cause seabirds to lose their sight.) The gular pouch inflates like a vacuum bag to cartoonish proportions, suctioning up water and hopefully netting the bird's next meal. Veteran divers snare fish about two-thirds of the time, while novices gradually master the art of diving through trial and error (fewer than half of all fledglings survive the learning curve; the rest succumb to starvation). Back at the surface, before swallowing her catch down, the pelican empties her water-filled pouch with a few tilts, shakes, and wobbles, a procedure that can take up to a minute and expel upwards of nearly three gallons of water. For good reason, French-speaking Louisianians nicknamed the brown pelican *grand gosier*—literally "big gullet."

Pelicans are what scientists call "opportunistic" or "generalist" feeders. Though they prefer small, finny fish, they might, whether purposefully or inadvertently, snack on some shrimp or slurp a venomous yellow-bellied sea snake. The world is their oyster—except they don't eat oysters—and they have been observed ingesting most anything that swims. Walter Anderson watched in awe as a Chandeleur pelican struggled to devour a full-grown terrapin. A pair of ornithologists found a California brown pelican that had died from trying to gulp down a stingray—the fish's spine had embedded itself in the bird's throat. Many more have asphyxiated while attempting to swallow invasive sailfin armored catfish off the coast of Puerto Rico. During periods of food scarcity and starvation, brown pelicans will eat to survive, and have been spotted preying upon the chicks of a cliff-dwelling seabird called the common murre, various egret species, and even their own young. Unfortunately, a pelican today is just as likely to swallow a mouthful of microplastics as a menhaden.

Recently, the internet has peddled the pelican as a most meme-worthy bird. Every year or so, social media lights up with videos of pelicans reaching their bills upward to stretch their prodigious pouches up, inside out, and over their own heads, giving the appearance that their spines are protruding from their mouths. Contrary to the accompanying clickbaity titles—my favorites include, "The Horrifying Act of Yawning Pelicans" and "A Pelican's Yawn Is Truly One of the Most Terrifying Things in Nature"—these acts of gular yoga simply allow pelicans to exercise their pouch muscles and thermoregulate, in a motion called gular fluttering.

But yawning pelicans pale in comparison to the number of internet slideshows and videos featuring pelican "attacks" from around the world. These images show birds not necessarily eating, but working their bills around—in an action we might term *pouching*—animals they probably shouldn't: cats, capybaras, other birds of various sizes. One popular YouTube video, gleefully titled "Pelican Swallows Ducklings 'OM NOM OM' (Mother Watches)," shows a Florida brown pelican playfully, or some might say sadistically, pouching mallard ducklings while a crowd looks on in horror.

These highly memorable (and meme-able) images offer an alternative to the long-held vision of the saintly, sacrificial pelican, that

Christlike creature. This reimagining reached its apotheosis with a 2012 article from the sports news site Deadspin titled, "Fuck You, Pelicans Are Awesome: A Defense of the NBA's Best New Team Name." Its author, Barry Petchesky, took issue with protests that New Orleans's professional basketball franchise had chosen a new name not formidable enough to challenge "fierce" mascots like the Raptors and Grizzlies on the court. Petchesky's article—written, assumably, with tongue planted firmly in cheek—argues that the pelican is not "a weak, floppy bird," as a rival sports journalist griped, but "fearsome." "The pelican is the serial killer of birds," Petchesky writes. "Not only is it a carnivore—it is a *hypercarnivore.*" (Which is definitionally true: its diet consists of more than 70 percent meat, but so does the cute and cuddly penguin's.) "What's more," he continues, "unlike an eagle or a falcon—the pelican almost never scavenges someone else's kill. It craves warm flesh, so it gets the job done itself." The pelican, Petchesky concludes, is a "badass."

* * *

On an early May morning in 2020, Travis Moore patrolled the sandy shores of Queen Bess Island, a significant brown pelican rookery near Grand Isle dating back to the 1970s translocation efforts. Working for the Coastal Protection and Restoration Authority, a Louisiana agency charged with safeguarding the state's coastline and wetlands, Moore had the job that day of numbering nests and nestlings in order to survey the health of the island's pelican population. At one point, a pelican thudded to a stop atop a mangrove thicket, not eight feet away from where Moore stood. He noticed that the bird sported a plastic, bubblegum-pink band on its left leg, and snapped a quick photo with his iPhone. That tag identified the bird as A04, an adult male pelican captured on August 3, 2010, just off the Mississippi coast, by rescuers collecting wildlife affected by the Deepwater Horizon disaster. A04 counted as one of the lucky 1,200-plus seabirds deemed healthy enough to survive rehab, lathered in Dawn dish detergent suds to remove all traces of oil, and eventually released.

In the photo, A04's blue-gray iris peers directly into Moore's lens. He stands tall, proud perhaps, and seemingly smiling, his breeding season plumage bursting with roses, golds, and rich browns. Two chicks, wine-dark, featherless, and not more than a few weeks old, helplessly grovel at their father's feet, begging for shade from the sun's ruthless rays.

A04's arrival and eventual brooding on Queen Bess, where the oil spill decimated not only a bird habitat but crucial oystering and fishing zones, provided a rare (and very real) "nature is healing" moment. News reports seized on the brown pelican's resilience—and, yes, maybe even the bird's badassery—while drawing connections to the inherently resilient nature of Louisianians. "A04 and the rebirth of Queen Bess Island are symbols of Louisiana's unwavering spirit," Governor John Bel Edwards wrote on Twitter. "When times are tough, we endure, and we come back stronger."

The governor's words were a telling reminder that Louisiana is a place constantly under siege by its intimate and often savage relationships with its own flora and fauna, people and petroleum, land and water. Due to centuries of environmental neglect and ecological destruction, Louisianians, like the Pelican in Her Piety that adorns their state flag, have learned to suffer, to sacrifice, to marry grief with compassion, even to bleed. But we've also learned to relish our suffering, to celebrate our sacrifices, to endure grief, to collectively bleed with compassion and joy and love for Louisiana. To avoid suffering is but a temporary convenience, Aesop's pelican tells the ostrich. Pain is a pleasure, she explains, "the most exquisite that nature hath indulged to us," a pleasure "in which pain itself is swallowed up and lost, [and] only serves to heighten the enjoyment."

5

The Pelican's Purpose

You try to fix a wild, hurt thing and what happens is, the same thing happens.
—Joy Williams, *Breaking and Entering*

We woke before dawn, with the intention of meeting the sun out at sea. But fog cloaked the Gulf Coast, rendering its waters unnavigable, at least for the next few hours. It was a typically contradictory week of shifting weather patterns in south Louisiana: freezing temperatures sandwiched between sweltering spells of humidity. The climate's inconsistency, paired with the coast's sinking landscapes and rising seascapes—what one writer of old called "the great winding world of liquid soil"—made it easy to feel disoriented.

I gazed out into the gloom to recalibrate. It was late February, ten days after Mardi Gras, and I was at the Louisiana Universities Marine Consortium, or LUMCON, a state-operated research and education campus located in the waterfront town of Cocodrie—Louisiana French

Opposite: Juita Martinez with pelican. Photograph by Liz Bourgeois.

for *crocodile*—where Highway 56 dead-ends into the Gulf of Mexico. The night before, I drove the two hours from New Orleans to join Juita Martinez, a pelicanologist in her fifth and final year of dissertation field research surveying three brown pelican nesting islands currently being enlarged and restored with funds from BP's $20.8 billion oil spill settlement, the largest environmental pollution payout in U.S. history ($403.9 million was specifically earmarked for bird conservation). The main goal of these coastal restoration projects, among many others, is to protect the human infrastructure that inhabits Louisiana's ever-receding coastline—a football field's worth of land every hundred minutes, as the oft-repeated statistic grimly states.

The fact that these restored islands are also seabird breeding grounds—potentially *better* breeding grounds than unrestored islands is Martinez's working hypothesis—is a fortuitous side effect, or as we say in French-accented south Louisiana, *lagniappe.* This is especially crucial considering that, in the decade following the oil spill, the number of Louisiana brown pelican colonies decreased by more than 50 percent.

When enough of the fog lifted, Martinez, her undergraduate assistant, and I packed a day's worth of gear into the *Camellia,* our LUMCON-furnished boat, and motored down Little Cocodrie Bayou, past marsh-dredged canals carved out decades ago by the oil industry; past ghostly cypress trees hollowed out by saltwater intrusion; past long-abandoned fishing camps collapsing into the brackish muck; past pelicans posing on lone pilings that once supported docks and piers since rotted or swept away; past fields of capped oil and gas wellheads—commonly called Christmas trees—that poke up from the water's surface; past groups of pelicans in single file, soaring their way seaward, successively flapping their wings down the line like Rockettes; into Terrebonne Bay, where the foghorns of hamlet-sized offshore oil platforms echo off the last slivers of land before the Gulf's vast, seemingly empty expanses; and finally to Raccoon Island.

This far from the coast, over twenty miles out from Cocodrie, the fog enveloped us once again, blotting out sea, sky, and sun like a purple bruise. Navigating blind if not for the boat's GPS, we lumbered toward Raccoon at a manatee's pace, following the sinister cackle of laughing gulls, like early explorers searching for land. "We should start to smell the island soon," Martinez shouted over the motor's roar into the haze. Pelican breeding grounds are notoriously odoriferous, gag-inducing isles of guano and regorged fish guts. I scented nothing but thought I could make out the faint blur of a distant shoreline. Suddenly, as if by epiphany, the sun melted the last remnants of morning mist and Raccoon Island exploded into view. "Pelicans!" I cheered, as a small pod levitated, circled, and relocated elsewhere at the sound of my voice. We anchored, gathered our gear, slipped into chest-high waders, and waddled the remaining hundred yards to shore.

Measuring a mile and a half in length, Raccoon Island is a 242-acre sliver of beaches, tidal marshes, and shrub-covered mini dunes. Geomorphically, Raccoon is both thousands of years old and relatively new. On August 10, 1856, a Category 4 hurricane pummeled Isle Dernière, or Last Island, a twenty-five-mile-long barrier island, resort, and gambling mecca known for having white sand beaches as "firm, compact, and smooth as chiseled marble." Over two hundred people perished in the storm, one of the strongest to ever strike

Louisiana, and Isle Dernière shattered into five fragments. Over time, the westernmost island of Raccoon became a nesting and migratory colony for brown pelicans and other waterbirds, including piping plovers, a near-threatened shorebird that winters here. With the island now under the jurisdiction of the Louisiana Department of Wildlife and Fisheries (LDWF) and off-limits to all but permitted visitors, engineers have, since restoration efforts began in 1997, pumped in sand dredged from the island's bay side, planted black mangrove and other soil-stabilizing species, and constructed a series of erosion-arresting breakwaters on the island's Gulf side—basically everything a viable pelican habitat would need.

Last year on Raccoon, Martinez witnessed an island overflowing with pelican parents—an estimated ten thousand couples. This year, we counted two dozen pelicans at most—probably the same stragglers we first spotted from the boat—scattered among the breakwaters. For two hours, we plodded, sun stunned, along the beaches and dunes of Raccoon, hoping to find nests. Except for the susurration of sand blowing through marsh grasses, the island remained silent. We found several pelican carcasses and bleached bones, young and old, including one recent set of remains, its feathers still neatly preened from its natural oils, that Martinez identified as a migrant red beard from California. Who knows how or why it flew this far.

We discovered a handful of deserted nests, no more than half a dozen. I plucked an abandoned egg from the ground and gently cradled the softball-sized ovoid in my hands, hoping to take it home. Like most brown pelican eggs, its chalk-white surface was streaked with blood. *It'll stink up your car for months,* Martinez warned.

"What happened?" I asked her. "Where are all the pelicans?"

It had to be the late freeze, she theorized. Cold spells send menhaden, the pelican's primary food source, swimming deeper, farther out into the Gulf. These birds have nothing to eat. And with nothing to eat they cannot bear eggs or, if they'd already laid a clutch, cannot bear the labor that incubation entails.

We slunk back to the *Camellia* and returned to LUMCON to refuel for our trip to the next island. There, the dock attendant asked if we'd seen the dying pelican by the boat launch. We hurried over to find the bird sitting still and sphinxlike, feet and pouch tucked into chest. *She's starving,* Martinez whispered. We mobilized without thinking. Though we might not know what happened to Raccoon Island's twenty thousand pelicans, we could still try to save this one. I volunteered to forgo the next island reconnaissance to ferry the bird to the Audubon Zoo's emergency wildlife clinic in New Orleans. Martinez's assistant salvaged a cardboard box—makeshift transport—from a nearby dumpster. We circled the frightened pelican as she began to flap her wings, but she was emaciated, unable to alight. I lunged, lightly pinning her to the asphalt. "Grab her beak!" Martinez cried. "Control the beak, control the pelican." I placed my left peace fingers between the bird's mandibles, wrapping my palm around her pouch. She ceased to struggle. With my right hand I held her wings closed while caressing her back—all inviting, velvety feathers. I daydreamed about laying my cheek upon her breast.

Martinez woke me from my reverie: "Lift her up and into the box." We sidestep-shuffled to my car, the caged bird between us. "Her name is Joval," I said, adopting the name of the manufacturing company printed on the side of the box. I cued up a playlist of soothing ocean sounds as Joval and I sped away from the coast.

Two days later, the zoo's senior veterinarian called to tell me that Joval did not survive.

* * *

It's easy to imagine how the brown pelican's end will arrive. Louisiana's population will survive these wintry aberrations, but they cannot outfly the gradually cataclysmic rising of the tides. California's Anacapa colony will survive in the island's high cliffs, but Florida's Pelican Island will silently succumb to the sea. Raccoon Island and all of Louisiana's barrier islands and coastal lowlands will disappear beneath the Gulf's murky waters. The pelicans know not what future awaits, but we keep pushing back against the most dire (and plausible) projections with each island habitat we restore.

One month following my trip to Raccoon, I met Martinez at Grand Isle, a quiet fishing community and Louisiana's last inhabited barrier island. The next day, blessed with ideal weather, we set off at dawn for Queen Bess Island, which a translocated pod of Florida pelicans brazenly colonized in 1973 to produce the first Louisiana-born nestlings in over a decade.

Today, the state's fourth largest pelican colony might as well be a different island. State authorities tripled the size of Queen Bess in the 1990s using sediment fill and an erosion-preventing rock containment ring around its perimeter. But the 2010 BP oil spill decimated the pelican colony. "You'd be hard pressed to find other areas that were oiled worse than Queen Bess," LDWF biologist Todd Baker told *National*

Geographic. Pictures taken several years after the spill show a land more liquid than solid, Gulf waters having leapt the neglected rock ring to swamp the island's interior, rendering only five of its thirty-six total acres viable for nesting. In August 2019, flush with BP cash, the state's Coastal Protection and Restoration Authority (CPRA), in partnership with the LDWF, launched a multifaceted $18.7 million restoration project that included a rebuilt rock barrier; the deposit of over 150,000 cubic yards of Mississippi River sediment; the planting of over 25,000 bushes and shrubs; and the installation of breakwaters, a seven-acre gravel beach for ground-nesting terns and black skimmers, and numerous structures called "bird ramps" that provide flightless juveniles with safe and easy access to the Gulf.

As I followed Martinez across Queen Bess, I was hard-pressed not to think of the proliferation of hypermodern, architecturally driven sports arenas—a Superdome dedicated to seabirds. The gravel beach, occupying the southwestern third of the horseshoe crab–shaped island, resembles nothing if not a sprawling parking lot, while the adjacent vegetative planting zone acts as a concession area of sorts for pelicans to forage for twigs and branches to build nests in zone three, a marshy, mangrove-rich breeding ground.

From a distance of fifty yards, I could see thousands of brooding brown pelicans perched atop mangroves, nestled between mangroves, in mangroves, under mangroves, hanging off the sides of mangroves at impossibly uncomfortable angles. We moved quickly but deliberately to create as little disturbance to the nests as possible. Yet no matter how quietly, how respectfully we approached the colony, one, then three, then a domino effect of dozens, and, eventually, hundreds of pelicans ascended with an audible *whup, whup, whup* of their wings to create a spectacular and more than slightly sinister, cloud-bursting, birdish sky.

Each bird had momentarily deserted a nest, an unmade bed of a pelican palimpsest filled with two eggs, sometimes three, which, if exposed for too long, would cook in the sun. I hastily helped Martinez switch out the batteries and data cards on ten motion-activated cameras she positioned around the colony to monitor nest activity. Within a month's time, those eggs would hatch wrinkly, featherless, purplish chicks that only a parent and Martinez could love—she calls them "dino floofs." Due to the cold spell, which caused a delay in the colony's nesting cycle, those chicks would be forced to fledge well beyond the start of hurricane season, when, under normal circumstances, juvenile birds would have already left the nest.

While Martinez reformatted a stubborn data card, I noticed a thick strand of bright yellow rigging rope knotted around the branches of a nearby mangrove, just below a cluster of egg-laden nests. Abandoned fishing tackle and other maritime detritus—ropes, hooks, and especially monofilament line—are far and away the number one manmade menace to brown pelicans. Throughout Florida and southern California, numerous wildlife rescue and rehabilitation centers spend most of their resources aiding ensnared and entangled pelicans. I pulled out my pocketknife and got to work.

As I hacked at the rope, my fingers slick with pale, mucilaginous pelican poop, I remembered a biological term given to wildlife that require humanity's interventional management in order to survive: "conservation-reliant" species. A better name, the environmental journalist Elizabeth Kolbert writes, might be "Stockholm species." For now, the brown pelican remains entirely dependent on the largesse of state and federal coffers, at least in Louisiana, where a seemingly unending stream of upcoming pelican projects continues to wind down the bureaucratic pipeline. The lives of bird and man remain tightly entwined like the strands of the rope I finally manage to pull free.

The CPRA's ten-year contract for the management and maintenance of Queen Bess expires in 2030, begging the question: What happens then? Do the pelicans wait for the next hurricane, the next oil spill, the next glacier-melt-causing centimetric rise in sea levels, the next undreamt cataclysmic event for humanity to step in and fix the problem all over again? What will become of the brown pelican? What will become of us? A bioindicator species, the brown pelican just might be the greater Gulf's canary on the coastline.

* * *

I returned to Queen Bess three months later, in June 2021. The island's nests now brimmed with baby pelicans, one- and two-month-old nestling pairs huddled together, thermoregulating their pale and pin-feathered bodies beneath the sun's midmorning sizzle. They regarded me warily as I tiptoed around the mangroves, shadowing Martinez. If I stepped an inch too close, the pelicans lunged with hiccupy and unmistakably dino-like squonks, out of fear or hunger—perhaps mistaking me for a parent returning to the nest with a face full of fish—I do not know.

I thought of Walter Anderson that morning—the artist sitting, sketching, living amongst the pelicans on his own island, seventy-five miles due northwest—as I watched these vulnerable baby birds cumbrously exercise their wings. I tried to grasp the full weight of what the re-disappearance of the brown pelican would mean to Louisiana's shores, what the loss of just one unfledged bird would mean to this tiny, sandy sliver in the sea, what Anderson's grief must have felt like when his colony of pelicans vanished.

Despite our best efforts, the brown pelican will not go extinct in our lifetime, nor the lifetimes of those who will best remember us. We will lose other species, other birds, too many to fathom, but the pelican will remain, for now. The species has survived over a century's worth of man's wars against nature, but the brown pelican will survive mankind, perhaps in diminished numbers, while likely losing historical nesting grounds to rising tides.

The "pelican holds everything," Anderson wrote nearly seventy-five years ago. A bird that could hold the world in its pouch meant everything in the world to the artist, and if we allow it, if we act as if we deserve it, the pelican can hold everything for us. For a species that often finds it hard to love his fellow man, we can find in pelicans a creature to love and can build a kinship to prepare ourselves for the litany of losses to come. "Why love what you will lose?" the poet Louise Glück once asked, before answering: "There is nothing else to love."

acknowledgments

I am grateful to the following individuals who made this book possible. Jenny Keegan and the LSU Press team were a pleasure to work with. Pelican scientists Betty Anne Schreiber Schenk, Robert Risebrough, Dan Anderson, and Brock Geary provided invaluable interviews. Travis Moore and Renee Bennett of the Coastal Protection and Restoration Authority answered questions about coastal restoration. Barret Fortier, Jimmy Laurent, and Taylor Pool of the U.S. Fish & Wildlife Service let me tag along on a trip to North Breton to survey the pelican colony. Lauren Leonpacher, outreach coordinator for the Coastal Wetlands Planning, Protection, and Restoration Act, helped with a query regarding the funding of Raccoon Island's restoration. Archivist extraordinaire Bobby Ticknor sourced key documents. Elizabeth Gross talked birds at a crucial moment. Aaron Clark-Rizzio informed me of a particularly important essay I had missed. Stephen Quinn provided an answer to the fate of the original Pelican Island diorama. Denny Culbert gifted a pelican photograph that hangs near my writing desk; its watchful eye acted as a reminder to do the work.

Liz Bourgeois contributed photos. Matt Smith encouraged me to go outside and look at birds. And big thanks to Susie Penman, who read and edited—for my fourth book in a row—the whole dang thing.

Finally, this book could not have been written without the guidance and support of Juita Martinez.

A portion of the proceeds from this book's sale will be donated to coastal restoration relief, wildlife rehabilitation efforts, and organizations dedicated to creating a more diverse and inclusive birding community. For more information, visit rienfertel.com.

notes

1. Familiars

1 Walter Inglis Anderson: On the life of Walter Anderson, see: Walter Anderson, *Birds* (Jackson: Univ. Press of Mississippi, 1990); Redding S. Sugg Jr., ed., *The Horn Island Logs of Walter Inglis Anderson,* rev. ed. (Jackson: Univ. Press of Mississippi, 1985); Christopher Maurer, *Fortune's Favorite Child: The Uneasy Life of Walter Anderson* (Jackson: Univ. Press of Mississippi, 2003); Walter Anderson, *Pelicans* (San Francisco: Cadmus Editions, 2004).

2 "the miracle of realizing that art": Anderson, *Pelicans,* 3.

2 "The whole island seems a concentration": Mary Anderson Pickard, "Preface," in Anderson, *Pelicans,* ix.

2 "When he saw a bird": Pickard, "Preface," v.

2 "his totem and a spiritual guide": Pickard, "Preface," vi.

3 "He might draw a hundred pelicans": Sugg, 33.

3 "all their reactions and conditions": Mary Anderson Pickard, "The Birds of Walter Anderson," in Anderson, *Birds,* xvi.

3 "tremendously musical harmonies": Pickard, "Preface," vi.

3 "to make myself a nest": Sugg, 48.

3 "Pelican Dictionary of Common Terms": Anderson, *Pelicans,* 28.

4 "taste good but the consistency": Sugg, 45.

4 "better raw than cooked": Sugg, 56.

4 "attacked several times": Sugg, 69.

4 "ecstasy of feeding": Sugg, 73.

4 "After you have lived on the island": Anderson, *Pelicans,* 1.

5   "it's impossible to get all": Sugg, 44.

6–7   The earliest known pelican fossil: Antoine Louchart, Nicolas Tourment, and Julie Carrier, "The Earliest Known Pelican Reveals 30 Million Years of Evolutionary Stasis in Beak Morphology," *Journal of Ornithology* 152 (2011): 15–20.

7   "living fossils": Brian Switek, "The Pelican's Beak: Success and Evolutionary Stasis," *Wired,* Sept. 20, 2010, https://www.wired.com/2010/09/the-pelicans-beak -success-and-evolutionary-stasis/.

7   "one of the most superficially dinosaur-like": Brett Westwood and Stephen Moss, *Natural Histories: 25 Extraordinary Species That Have Changed Our World* (London: John Murray, 2015), 142.

7   "earlier and more primitive forms": Pickard, "The Birds of Walter Anderson," xvi.

7   "prehistoric monsters": Sugg, 220.

7   "In a word . . . you lose": Anderson, *Pelicans,* 1.

8   "Providence . . . has been kind": Sugg, 213.

8   "Three pelicans—then six pelicans": Sugg, 220.

8   "Pelicans! . . . seventeen in one flock!": Sugg, 223.

8   "Pollution now is one of the most": Lyndon B. Johnson, "Statement by the President in Response to Science Advisory Committee Report on Pollution of Air, Soil, and Waters," Nov. 6, 1965, https://www.presidency.ucsb.edu/documents/statement -the-president-response-science-advisory-committee-report-pollution-air-soil-and.

9   "'hope' is the thing with feathers": *The Complete Poems of Emily Dickinson,* ed. Thomas H. Johnson (New York: Back Bay Books, 1976), 116.

2. Pelicanland

11   five blue jays: Frank M. Chapman, "Birds and Bonnets," *Forest and Stream,* Feb. 25, 1886, 84.

12   Entire nature tableaux spilled forth: Jennifer Price, "Hats Off to Audubon," *Audubon,* Nov./Dec. 2004, 44–50.

12   At the carnage's height: Frank B. Gill, *Ornithology,* 2nd ed. (New York: W. H. Freeman and Co., 1995), 601–2.

12   "slaughter ground of the plume and wing hunters": Alfred M. Bailey, "Observations on the Water Birds of Louisiana," *Natural History* 19 (1919): 44–56.

13   "The land which produced Audubon": [George Bird Grinnell,] "The Audubon Society," *Forest and Stream,* Feb. 11, 1886, 41.

13   "dead bird wearing gender": J. A. Allen, "The Audubon Society," *Forest and Stream,* Mar. 4, 1886, 103.

13   "The reform in America, as elsewhere": [Grinnell,] "The Audubon Society."

13 Audubon societies soon sprung up nationwide: Carolyn Merchant, "George Bird Grinnell's Audubon Society: Bridging the Gender Divide in Conservation," *Environmental History* 15 (Jan. 2010).

14 "We flattered ourselves": Celia Thaxter, "Woman's Heartlessness," *Audubon Magazine*, Feb. 1887, 13–14.

15 "I was anxious to kill": John James Audubon, "A Naturalist's Excursion in Florida," *Supplement to the Connecticut Courant*, May 15, 1832, 41–44.

15 "Good luck to you poor doomed creatures": Mark V. Barrow Jr., *A Passion for Birds: American Ornithology after Audubon* (Princeton: Princeton Univ. Press, 1998), chapter 5.

15 "lady of the stupid face": Virginia Woolf, "The Plumage Bill," *Woman's Leader*, July 23, 1920, 559–60.

16 Boston socialite cousins: Price, "Hats Off to Audubon."

16 "quills to avoid": M. O. W. [Mabel Osgood Wright], "Consistency," *Bird-Lore*, Oct. 1899, 170–72.

17 "That there should be an owl or ostrich": Price, "Hats Off to Audubon."

17 A rich reddish brown with white-to-silvery-gray highlights: M. O. W., "Consistency."

17 "The feathers of this bird are now worn so commonly": [Frank M. Chapman,] "Editorials," *Bird-Lore*, Oct. 1899, 169.

17 "lazy": L. W. Brownall, "A Visit to Pelican Island, on Indian River, Florida," *The Osprey*, Jan. 1899, 70–71.

17 "heavy and clumsy": Dr. J. B. Holder, "The Brown Pelican and Its Home," *American Sportsman*, Mar. 21, 1874, 390.

17 "Homely": George H. Lowery Jr., *Louisiana Birds*, 3rd ed. (Baton Rouge: Louisiana State Univ. Press, 1974), 128.

17 "Graceless": Bill Thomas, "The Brown Pelican: A Wonderful Bird Indeed—But a Frightening Omen," *Defenders*, Dec. 1976, 362–67.

17 "Gluttonous": George S. Fichter, "A Peculiar Bird Is the Pelican," *American Mercury*, Dec. 1956, 69–74.

17 "Gawky": Paul Stoutenburgh, "The Pelican's Gawky Grace," *Suffolk Times*, May 5, 2005.

17 "Ludicrous": John W. Daniel Jr., "Pelicans of Tampa Bay," *Wilson Bulletin* 14:1 (Mar. 1902): 5–7.

17 "Queer": Lowery, *Louisiana Birds*, 128.

17 "Strange and comical": Thomas, "The Brown Pelican: A Wonderful Bird Indeed."

17 "Strange, weird creatures" and "Such peculiar birds": Edward Howe Forbush and John Bichard May, *A Natural History of American Birds of Eastern and Central North America* (1925; New York: Bramhall House, 1955), 20.

17 "Awkward, ungainly birds": Daniel, "Pelicans of Tampa Bay."

17 "Silent, dignified and stupid birds": Arthur Cleveland Bent, *Life Histories of North American Petrels and Pelicans and Their Allies* (Washington D.C.: Smithsonian Institution, United States National Museum, Bulletin 121, 1922), 295.

17 "This tragicomic bird" and "This comedian of the seas": Ralph W. Schreiber, "Bad Days for the Brown Pelican," *National Geographic,* Jan. 1975, 110–23.

17 "Extremely stupid": Charles F. Holder, "Interesting Facts about Pelicans," *Scientific American,* June 27, 1903, 489.

17 "Stately and solemn": Fichter, "A Peculiar Bird Is the Pelican."

17 "Stately but slow-witted": George Laycock, "Where Have All the Pelicans Gone?" *Audubon,* Sept. 1969, 10–17.

17 "sad melancholy expression": Hannah Yates, *Official Louisiana Bird: The Brown Pelican* (New Orleans: Louisiana Wildlife and Fisheries Commission [1962]), 2.

17 "Pelicans . . . are not emotional birds": Editorial, *Bird-Lore,* May–June 1924, 207.

18 "I know of no more magnificent sight": Bent, *Life Histories of North American Petrels and Pelicans,* 289.

18 "incongruous and grotesque": Holder, "The Brown Pelican and Its Home."

18 "Grotesque fishermen": Bailey, "Observations on the Water Birds of Louisiana."

18 "grotesque and striking features": Bent, *Life Histories of North American Petrels and Pelicans,* 301.

18 "The least attractive": Holder, "Interesting Facts about Pelicans."

18 "The ugliest and most awkward": Brownall, "A Visit to Pelican Island."

18 "A disagreeable, wheezing, asthmatic bird": Holder, "Interesting Facts about Pelicans."

18 Frank Chapman: On Frank Chapman's dioramas, see: Stephen Christopher Quinn, *Windows on Nature: The Great Habitat Dioramas of the American Museum of Natural History* (New York: Abrams, 2006), 15–18, 79–85.

19 "Pelicanland": Frank M. Chapman, *Camps and Cruises of an Ornithologist* (New York: D. Appleton and Company, 1908), 92.

19 President Theodore Roosevelt: On Roosevelt's love for birds and quotations, see: Duane G. Jundt, "'I So Declare It': Theodore Roosevelt's Love Affair with Birds" in Char Miller and Clay S. Jenkinson, eds., *Theodore Roosevelt: Naturalist in the Arena* (Lincoln: Univ. of Nebraska Press, 2020), 45–63; for his White House life list, see: "President Roosevelt's List of Birds," *Bird-Lore,* Mar.–Apr. 1910, 53–55.

20 "I need hardly say how heartily": "A Letter from Governor Roosevelt," *Bird-Lore,* Apr. 1899, 65.

21 "To me they are always interesting and amusing": Theodore Roosevelt, "Harpooning Devilfish," *Scribner's Magazine,* Sept. 1917, 293–305.

22 "Remain there . . . till your heart is glutted": Robert Barnwell Roosevelt, *Florida and the Game Water-Birds* (New York: Orange Judd Company, 1884), 96.

22 "I found them breeding in larger and larger numbers": Henry Bryant, "January 19, 1859," *Proceedings of the Boston Society of Natural History* 7:1 (Mar. 1859): 18–20.

22 "The mangroves and water-oaks of this island": James A. Henshall, *Camping and Cruising in Florida* (Cincinnati: Robert Clarke & Co., 1884), 57.

23 "adhere to a chosen site": Morris Gibbs, "Nesting Habits of the Brown Pelican in Florida," *The Oölogist* 11:3 (Mar. 1894): 81–84.

23 "No traveler ever entered the gates": Frank M. Chapman, *Bird Studies with a Camera* (New York: D. Appleton and Co., 1900), 191–214.

24 "brutally persecuted": Frank M. Chapman, "Pelican Island Revisited," *Bird-Lore,* Jan.–Feb. 1901, 3–8.

24 McIlhenny nurtured eight snowy egrets: E. A. McIlhenny, *Bird City* (Boston: Christopher Publishing House, 1934).

25 "Is there any law that will prevent": Douglas Brinkley, *The Wilderness Warrior: Theodore Roosevelt and the Crusade for America* (New York: Harper, 2009), 13–14.

25 "To see a file of pelicans winging": *Roosevelt, Friend of the Birds* (Roosevelt Memorial Association Film Library, 1924).

25 Paul Kroegel: For biographical information on Paul Kroegel, see: Brinkley, *The Wilderness Warrior,* chapter 18; George Laycock, "Where Have All the Pelicans Gone?" *Audubon,* Sept. 1969, 10–17; "A Brief Biography of Kroegel," U.S. Fish and Wildlife Service, Mar. 7, 2014, https://training.fws.gov/History/ConservationHeroes/Kroegel.html.

27 "all the activities of Pelican home-life": [Frank M. Chapman,] "Editor's Notes," *Bird-Lore,* Mar.–Apr. 1905, 148.

27 "Ceremony of Nest Relief": Chapman, *Camps and Cruises of an Ornithologist,* 95.

27 Guy Bradley: On Guy Bradley and Columbus G. McLeod, see: "Game Warden Guy M. Bradley," Officer Down Memorial Page, https://www.odmp.org/officer/2150-game-warden-guy-m-bradley; "Game Warden Columbus G. McLeod," Officer Down Memorial Page, https://www.odmp.org/officer/24122-game-warden-columbus-g-mcleod.

28 "It has been a question with me": Gibbs, "Nesting Habits of the Brown Pelican in Florida."

28 "The claim is being put forth": "Death to the Pelican!" *Bird-Lore,* Mar.–Apr. 1918, 194–98.

29 "Kill the pelican": T. Gilbert Pearson, "The Case of the Brown Pelican," *American Review of Reviews* 59:5 (May 1919): 509–11.

29 "it is time that the Government was informed": "Death to the Pelican!"

29 The deputy fish commissioner of St. Petersburg: "Death to the Pelican!"

29 In New Orleans, at a meeting: Florida Audubon Society, *A Defense of the Pelican* (1918).

30 "the Pelican serves no useful purpose": "Death to the Pelican!"

30 Legislative bodies targeted other avian "pests": Pearson, "The Case of the Brown Pelican."

30 Yellowstone National Park rangers killed hundreds: Lynn M. Stone, *North American Pelicans* (Minneapolis: Carolrhoda Books, 2003), 40.

30 "A new carnival of bird slaughter": Florida Audubon Society, *A Defense of the Pelican.*

30 The government enlisted: Pearson, "The Case of the Brown Pelican." See also Alfred M. Bailey, "The Brown Pelican: A Good Citizen," *Wilson Bulletin* 30:3 (Sept. 1918): 65–68.

31 Pelicans, and other sea surface–feeding fowl, actually help: "In Defense of the Pelican," *Bulletin of the Massachusetts Audubon Society* 2:3 (Apr. 1918): 4.

31 members of the Halifax River Bird Club: "Slaughter of Brown Pelicans," *Bird-Lore,* May–June 1924, 221–22, 230.

32 "On a per-hectare basis, mangroves": María Verza, Christina Larson, and Victoria Milko, "Restoring Mexico's Mangroves Can Shield Shores, Store Carbon," *AP News,* Nov. 5, 2021.

3. No Oases of Safety

35 The Indigenous Chumash people: Erwin G. Gudde, *California Place Names: The Origin and Etymology of Current Geographical Names,* 4th ed. (Berkeley: Univ. of California Press, 1998), 12.

35 Most importantly, Anacapa hosts: Ralph W. Schreiber and Robert L. DeLong, "Brown Pelican Status in California," *Audubon Field Notes* 23 (1969): 57–59.

36 consider the infamous prison island: Gudde, 7.

36 "suddenly halted in shocked dismay": Joseph R. Jehl Jr., "A Wonderful Bird *Was* the Pelican," *Oceans* 2:3–4 (Sept.–Oct. 1969): 10–19.

37 Of the colony's 298 nests: Robert W. Risebrough et al., "Reproductive Failure of the Brown Pelicans on Anacapa Island in 1969," *American Birds* 25:1 (Feb. 1971): 8–9.

37 The first signs of trouble had appeared: Lowery, *Louisiana Birds,* 121–28.

38 "The ground was covered with eggs": Alfred M. Bailey, "The Brown Pelican: A Good Citizen," *Wilson Bulletin* 30:3 (Sep. 1918): 65–68.

38 pelican counters revisited the Chandeleurs' North Island: Lowery, 122–23.

39 Fishponds were emptied: Rachel Carson, *Silent Spring* (1962; Boston: Mariner Books, 2002), 140.

39 By 1963, an estimated five million: Lowery, 125.

39 "When we left Venice to run": Bob Marshall, "Pelicans: A Never-Ending Battle," *Times-Picayune,* May 4, 1988.

39 "blank spots": Jehl, "A Wonderful Bird Was the Pelican."

39   Towns throughout the country reported: Carson, *Silent Spring,* 167.

39   Robins dropped dead: Carson, 108.

39   The bobwhite quail and wild turkey: Carson, 167.

39   In Texas, sows birthed entire litters: Carson, 168.

39   Nursing calves died after: Carson, 168.

39   DDT exposure caused gynandromorph: Carson, 212.

39   In Pacific Canada, coho salmon: Carson, 136.

39   "The 'harmless' shower hath killed": Olga Owens Huckins to the Editor of the [Boston] *Herald,* Jan. 28, 1958, Duxbury, Mass., Rachel Carson Papers, YCAL MSS 46, Yale Univ. Library.

39   Huckins penned a second letter to her friend: Olga Huckins to Rachel [Carson], Jan. 27, 1958, Duxbury, Mass., Rachel Carson Papers, YCAL MSS 46, Yale Univ. Library.

40   "insect 'bomb'": "Science: War on Insects," *Time,* Aug. 27, 1945, 65.

40   DDT: On DDT, see: Carson, especially 18–23; Robert W. Risebrough et al., "DDT Residues in the Pacific Sea Birds: A Persistent Insecticide in Marine Food Chains," *Nature,* Nov. 11, 1967, 589–91; Daniel W. Anderson et al., "Significance of Chlorinated Hydrocarbon Residues to Breeding Pelicans and Cormorants," *The Canadian Field-Naturalist* 83:2 (Apr.–June 1969): 91–112; Michael Lipske, "How Rachel Carson Helped Save The Brown Pelican," *National Wildlife,* Dec. 1, 1999, https://www.nwf.org/Magazines/National-Wildlife/2000/How-Rachel-Carson-Helped-Save-The-Brown-Pelican.

41   "DDT presumably could send": Elena Conis, "Beyond Silent Spring: An Alternative History of DDT," *Distillations,* Feb. 14, 2017, https://www.sciencehistory.org/distillations/magazine/beyond-silent-spring-an-alternate-history-of-ddt.

41   "This is a book about man's war": Linda Lear, *Rachel Carson: Witness for Nature* (New York: Henry Holt, 1997), 386.

41   "agents of death": Carson, 18.

41   Endrin ranked as the most toxic: Carson, 26–27.

41   "It is not possible to add": Carson, 42.

41   "why a spinster with no children": Lear, *Rachel Carson: Witness for Nature,* 429.

41   "more poisonous than the pesticides": Lear, 432.

42   "biological magnification": Jehl, "A Wonderful Bird Was the Pelican."

42   DDT residue in human breast milk: Carson, 23.

42   "poisoned environment" and "no oases of safety remain": Carson, 87.

42   The age of the Anthropocene: Erle C. Ellis, *Anthropocene: A Very Short Introduction* (Oxford: Oxford Univ. Press, 2018), 152–53.

43   "We are interested in exploring": Dan Guravich and Joseph E. Brown, *The Return of the Brown Pelican* (Baton Rouge: Louisiana State Univ. Press, 1983), 54.

43   "Why not do something about it?": George Laycock, "Where Have All the Pelicans Gone?" *Audubon,* Sept. 1969, 10–17.

43   the "Pelican Committee": On the Pelican Committee, see: Guravich and Brown, *The Return of the Brown Pelican,* chapter 3; Laycock, "Where Have All the Pelicans Gone?"

46   "There's a big difference in the strength": Guravich and Brown, 56.

46   *Science* magazine published a paper: Joseph J. Hickey and Daniel W. Anderson, "Chlorinated Hydrocarbons and Eggshell Changes in Raptorial and Fish-Eating Birds," *Science,* Oct. 11, 1968, 271–73.

47   Anacapa eggshells measured at half the thickness: Daniel W. Anderson and Joseph J. Hickey, "Oological Data on Egg and Breeding Characteristics of Brown Pelicans," *Wilson Bulletin* 82:1 (Mar. 1970): 14–28; see also D. A. Ratcliffe, "Decrease in Eggshell Weight in Certain Birds of Prey," *Nature,* July 8, 1967, 208–210.

48   The Montrose Chemical Corporation: On the Montrose Chemical Corporation, see: Molly Peterson, "Your Local Superfund Site: Palos Verdes Shelf & Its Fish," *Pacific Swell,* June 3, 2011, https://www.scpr.org/blogs/environment/2011/06/03/3027 /know-your-local-superfund-site-palos-verdes-shelf-/; Rosanna Xia, "L.A.'s Coast Was Once a DDT Dumping Ground. No One Could See It—Until Now," *Los Angeles Times,* Oct. 25, 2020.

48   a recently released report blames: Rosanna Xia, "Sea Lions Are Dying from a Mysterious Cancer. The Culprits? Herpes and DDT," *Los Angeles Times,* Jan. 31, 2021.

48   the USDA had canceled the application: US Environmental Protection Agency, "DDT Regulatory History: A Brief Survey (to 1975)," July 1975, https://archive.epa.gov /epa/aboutepa/ddt-regulatory-history-brief-survey-1975.html.

49   "threatened with extinction": Ninety-second Cong., Sec. Sess., *Endangered Species Conservation Act of 1972,* 108.

49   The winter of 1971 brought: Guravich and Brown, 57; Cornelia Carrier, "40 Nests, Banded Birds: A Pelican Success Story," *Times-Picayune,* May 5, 1972; Stephen A. Nesbitt et al., "Brown Pelican Restocking Efforts in Louisiana," *Wilson Bulletin* 90:3 (Sept. 1978): 443–45.

50–51 Anacapa Island averages around 5,000: "California Brown Pelican," Channel Islands: National Park California website, https://www.nps.gov/chis/learn/nature/brown -pelican.htm.

51   "At a time when so many species of wildlife": David Max Braun, "Brown Pelican Off the Endangered Species List," *National Geographic Society Newsroom,* Nov. 14, 2009, https://blog.nationalgeographic.org/2009/11/14/brown-pelican-off-the -endangered-species-list/.

51   In late 1982, at Dana Point: "More Mutilated Pelicans," *UPI Archives,* Nov. 22, 1982, https://www.upi.com/Archives/1982/11/22/More-mutilated-pelicans/98714067

89200; Barbara Whitaker, "Killing and Maiming of Imperiled Pelicans Stun Officials," *New York Times,* Jan. 23, 2003.

51 The following year, at nearby Redondo Beach: "Two Bait Barge Workers Have Pleaded Innocent to 60," *UPI Archives,* Feb. 4, 1984, https://www.upi.com/Archives/1984/02/04/Two-bait-barge-workers-have-pleaded-innocent-to-60/7054444718800.

51 In Huntington Beach the same year: Frank Messina, "Pelican Mutilations Return with Vengeance," *Los Angeles Times,* Sep. 20, 1992.

51 Farther north in Monterey: "Pelicans Slashed on Monterey Bay," *UPI Archives,* Sept. 30, 1984, https://www.upi.com/Archives/1983/09/30/Pelicans-slashed-on-Monterey-Bay/4728433742400.

51 Scientists soon blamed El Niño: Messina, "Pelican Mutilations Return with Vengeance."

52 Every few years, especially in Florida: "Brown Pelicans Found with Pouches Slashed in Florida Keys," NBC 6 South Florida website, Dec. 12, 2013, https://www.nbcmiami.com/news/local/brown-pelicans-found-with-pouches-slashed-in-florida-keys/2014684; "Reward Offered for Pelican Throat-Slashing Culprits," WPLG Local 10 Miami website, Feb. 24, 2015, https://www.local10.com/news/2015/02/24/reward-offered-for-pelican-throat-slashing-culprits.

52 A decade following the first wave: Messina, "Pelican Mutilations Return with Vengeance."

4. Insatiable Creatures

57 Deepwater Horizon: For the Deepwater Horizon oil spill and its effects on birdlife, see: Scott Weidensaul, "Ten Years Later," *Living Bird* 39:4 (Autumn 2020): 18–33; J. V. Remsen et al., "The Regional, National, and International Importance of Louisiana's Coastal Avifauna," *Wilson Journal of Ornithology* 131:2 (June 2019): 221–42; J. Christopher Haney, Harold J. Geiger, and Jeffrey W. Short, "Bird Mortality from the Deepwater Horizon Oil Spill. I. Exposure Probability in the Offshore Gulf of Mexico" and "II. Carcass Sampling and Exposure Probability in the Coastal Gulf of Mexico," *Marine Ecology Progress Series* 513 (Oct. 22, 2014): 225–37, 239–52.

58 "spread a deathtrap over the waters": untitled editorial, *Bird-Lore,* Mar.–Apr. 1921, 101.

59 "The doorwings of the heavens": Barbara Allen, *Pelican* (London: Reakton, 2019), 61.

59 the Torah describes the pelican: Allen, *Pelican,* 62.

60 "What accident has befallen you": Monsieur de Meziriac, *Aesop's Fables: Together with the Life of Aesop* (Chicago: Rand, McNally & Company, 1897), 177–78.

61 "The pelican . . . is an exceeding lover": Michael J. Curley, trans., *Physiologus: A Medieval Book of Nature Lore* (Chicago: Univ. of Chicago Press, 2009), 9–10.

62 "The serpent hateth kindly this bird": Bartholomaeus Anglicus, *De proprietatibus rerum,* Book 12, trans. Robert Steele, http://bestiary.ca/beasts/beast244.htm.

62 "Never did mother-sheep love her lamb": Guillaume le Clerc, *Bestiaire,* trans. L. Oscar Kuhns, http://bestiary.ca/beasts/beast244.htm.

62 *"il nostro Pelicano"*: Victor E. Graham, "The Pelican as Image and Symbol," *Revue de littérature comparée,* April 1, 1962, 235–43.

62 "Lord Jesus, Good Pelican": Thomas Aquinas, *"Adoro Te Devote*—Hymn of Thomas Aquinas—Traditional Latin Gregorian Chant," YouTube video, https://youtu.be /q8rMi-DWT4k.

63 "when we concern ourselves less": *The Aberdeen Bestiary,* Special Collections and Museums, University of Aberdeen, https://www.abdn.ac.uk/bestiary/ms24/f35r.

63 "In the world there will always be men": Colin Thiele, *Storm Boy and Other Stories* (London: New Holland, 2018), 98.

64 "pelican daughters": William Shakespeare, *King Lear,* in *Shakespeare's Tragedies* (London: George Newnes, 1901), 681.

64 "To his good friends thus wide": William Shakespeare, *Hamlet,* in *Shakespeare's Tragedies* (London: George Newnes, 1901), 606.

64 "a creation which I, like the pelican": John Oxenford, trans., *Conversations of Goethe with Eckermann and Soret* (London: George Bell & Sons, 1882), 52.

64 "sacrifice divine": Alfred de Musset, "La Nuit de Mai," Poems Without Frontiers, http://www.poemswithoutfrontiers.org/La_Nuit_de_Mai.html.

64 "Would you know why I have done": *The Complete Works of Lord Byron,* ed. Thomas Moore, 2nd ed. (Frankfurt, Germany: Joseph Baer, 1852), 399.

65 five basic principles of flag design: Edward B. Kaye, "*Good Flag, Bad Flag,* and the Great NAVA Flag Survey of 2001," *Raven: A Journal of Vexillology* 8 (2001): 11–38, https://nava.org/digital-library/raven/Raven_v08_2001_p011–038.pdf.

66 Louisiana's pennant: On the history of the Louisiana seal and flag, see: Glen Duncan, *A Modern History of the Louisiana Pelican Flag: Or, a Tale of the Surprisingly Difficult Quest for the "Official" State Flag* (Baton Rouge: Self-published, 2010).

66 "The people of the new State have strange ideas": Henry L. Favrot, "The State Seal," *Publications of the Louisiana Historical Society* 2:4 (Dec. 1901): 21.

67 "especially during a time like this": Jeremy Alford, "Houma Teen Aims to Fix Louisiana Flag," *Houma Courier,* Apr. 6, 2006.

68 "an appropriate display of three drops": Duncan, 12.

68 "Our brown pelican is white": Smiley Anders, "Our Brown Pelican Is White," *The Advocate,* Feb. 14, 2017, https://www.theadvocate.com/baton_rouge/entertainment_life /smiley_anders/article_7fe30308-f225-11e6-bd0e-670f79c0c13f.html.

68   "A wonderful bird is the pelican": "D. L. Merritt, Wrote Limerick on Pelican," *New York Times,* Jan. 11, 1972, 40.

69   "If your food is all finned": Charley Harper, *Birds and Words* (Los Angeles: AMMO Books, 2008), 142.

69   "Pelicans that live beside rivers swallow": Aristotle, *The History of Animals,* Book 9, trans. D'Arcy Wentworth Thompson, https://penelope.uchicago.edu/aristotle/histanimals9.html.

69   "insatiable creatures": Pliny the Elder, *Natural History,* Book 10, http://bestiary.ca/beasts/beast244.htm.

70   "literally stuffed to the gills": George S. Fichter, "A Peculiar Bird Is the Pelican," *American Mercury,* Dec. 1956, 69–74.

70   the only member of the Pelecanus family that plunge dives: For the feeding habits of the brown pelican, see: Gordon H. Orians, "Age and Hunting Success in the Brown Pelican (*pelecanus occidentalis*)," *Animal Behavior* 17 (1969): 316–19; Ralph W. Schreiber et. al., "Prey Capture by the Brown Pelican," *The Auk* 92:4 (Oct. 1975): 649–54; Göran Arnqvist, "Brown Pelican Foraging Success Related to Age and Height of Dive," *The Condor* 94 (1992): 521–22; C. Claiborne Ray, "Pelicans in Training," *New York Times* online, Oct. 25, 2010, https://www.nytimes.com/2010/10/26/science/26qna.html; Sara Sneath, "How the Pelican State Nearly Lost Its Pelicans: Saving the Southern Wild," *Times-Picayune,* Aug. 23, 2018, https://www.nola.com/archive/article_4799dc0c-e5f4-5bfb-b403-9ec77c80d01d.html.

71   *grand gosier*: Joseph E. Brown, *The Return of the Brown Pelican* (Baton Rouge: Louisiana State Univ. Press, 1983), 5.

71   Walter Anderson watched in awe: Redding S. Sugg Jr., ed., *The Horn Island Logs of Walter Inglis Anderson,* rev. ed. (Jackson: Univ. Press of Mississippi, 1985), 75–76.

71   A pair of ornithologists found: Dennis L. Bostic and Richard C. Banks, "A Record of Stingray Predation by the Brown Pelican," *The Condor* 68:9 (Sept.–Oct. 1966): 515–16.

71   Many more have asphyxiated while: Lucy Bunkley-Williams et al., "The South American Sailfin Armored Catfish, *Liposarcus multiradiatus* (Hancock), a New Exotic Established in Puerto Rican Fresh Waters," *Caribbean Journal of Science* 30:1–2 (1994): 90–94.

72   During periods of food scarcity and starvation: Miguel A. Mora, "Predation by a Brown Pelican at a Mixed-Species Heronry," *The Condor* 91:3 (Aug. 1989): 742–43.

72   my favorites include: "The Horrifying Act of Yawning Pelicans," YouTube video, https://youtu.be/m76kL9VHrvE; Naveen A., "A Pelican's Yawn Is Truly One of the Most Terrifying Things in Nature," *Inner Splendor,* May 31, 2020, https://apsari.com/did-you-know-that-when-pelicans-yawn-it-looks-like-their-spine-is-coming-out-of-their-mouths.

72 One popular YouTube video: "Pelican Swallows Ducklings 'OM NOM OM' (Mother Watches)," YouTube video, https://youtu.be/dlJk0w4SnH8.

73 "The pelican is the serial killer of birds": Barry Petchesky, "Fuck You, Pelicans Are Awesome: A Defense of the NBA's Best New Team Name," Deadspin, Dec. 6, 2012, https://deadspin.com/fuck-you-pelicans-are-awesome-a-defense-of-the-nbas-b-5966336.

74 That tag identified the bird as A04: On A04 and Queen Bess, see: Irene Taylor Brodsky, dir., *Saving Pelican 895* (Vermilion Films: 2011); Emma Bryce, "How to De-Oil a Bird," Audubon website, May 4, 2015, https://www.audubon.org/news/how-de-oil-bird; LouisianaCPRA, @LouisianaCPRA, Twitter, May 5, 2020, https://twitter.com/LouisianaCPRA/status/1257761906435260422; Johnna Crider, "Healing from the Deepwater Horizon Oil Spill a Decade Later," *CleanTechnica*, May 6, 2020, https://cleantechnica.com/2020/05/06/healing-from-the-deepwater-horizon-oil-spill-a-decade-later; "Rescued from Deepwater Horizon, a Resilient Native Returns to Queen Bess Island," Gulf Spill Restoration website, June 10, 2020, https://www.gulfspillrestoration.noaa.gov/2020/06/rescued-deepwater-horizon-resilient-native-returns-queen-bess-island; Weidensaul, "Ten Years Later," 18–33.

74 one of the lucky 1,200-plus seabirds: Bob Johns, "Aftermath: The Gulf Oil Spill," *Bird Watcher's Digest*, https://www.birdwatchersdigest.com/bwdsite/solve/conservation/thegulfoilspill.php.

74 "A04 and the rebirth of Queen Bess": John Bel Edwards, @LouisianaGov, Twitter, May 5, 2020, https://twitter.com/LouisianaGov/status/1257768292279861257.

75 "the most exquisite that nature hath indulged": de Meziriac, *Aesop's Fables*, 178.

5. The Pelican's Purpose

78 Juita Martinez: On Juita Martinez's research, see: Scott Weidensaul, "Ten Years Later," *Living Bird* 39:4 (Autumn 2020): 18–33.

78 BP's $20.8 billion oil spill settlement: Rene Ebersole, "Critical Pelican Nesting Ground Restored, 10 Years after BP Oil Spill," *National Geographic* online, Apr. 6, 2020, https://www.nationalgeographic.com/animals/article/pelicans-queen-bess-island-restoration-gulf-oil-spill.

78 $403.9 million was specifically earmarked: "Plan for Deepwater Horizon Oil Spill Natural Resource Injury Restoration: An Overview," Apr. 2016, https://www.gulfspillrestoration.noaa.gov/sites/default/files/wp-content/uploads/Overview_04-07-16_final-508.pdf, 38.

78 the number of Louisiana brown pelican colonies decreased: Coastal Protection and Restoration Authority, "Queen Bess Island Restoration," Jan. 2020, http://coastal.la.gov/wp-content/uploads/2020/01/BA-0202-Queen-Bess-Island-Restoration-Fact-Sheet.pdf.

79   "firm, compact, and smooth": Bill Dixon, *Last Days of Last Island: The Hurricane of 1856, Louisiana's First Great Storm* (Lafayette: Univ. of Louisiana at Lafayette Press, 2009), 19.

80   engineers have, since restoration : "Barrier Island Status Report: Draft Fiscal Year 2021 Annual Plan," Jan. 2020, http://coastal.la.gov/wp-content/uploads/2019/01/AppB -FY21-Barrier-Island-Status-Report.pdf.

82   Pelican Island will silently succumb to the sea: Sustainability Leadership Service Learning Project, "Regional Sea Level Rise, Climate Change, and Species Adaptation Scenarios for Florida" (Norfolk, Va.: Old Dominion Univ., 2017), http://www.mari-odu .org/academics/2017su_leadership/documents/IDS496SustainabilityLeadership JointDocument.pdf.

82   the state's fourth largest pelican colony: National Oceanic and Atmospheric Ad- ministration, "Queen Bess Island Restored in Time for Nesting Season," Feb. 14, 2020, https://www.gulfspillrestoration.noaa.gov/2020/02/queen-bess-island -restored-time-nesting-season.

82   State authorities tripled the size of Queen Bess: Coastal Protection and Restoration Authority, "Queen Bess Island Restoration."

82   "You'd be hard pressed to find other areas": Ebersole, "Critical Pelican Nesting Ground Restored."

83   Pictures taken several years after the spill: Coastal Protection and Restoration Au- thority, "Queen Bess," Mar. 7, 2020, https://coastal.la.gov/news/queen-bess.

83   flush with BP cash: Ebersole, "Critical Pelican Nesting Ground Restored"; Coastal Protection and Restoration Authority, "Queen Bess Island Restoration."

84   "conservation-reliant" species: J. Michael Scott, "Recovery of Imperiled Species under the Endangered Species Act: The Need for a New Approach," *Frontiers in Ecology and the Environment* 3:7 (Sept. 2005): 383–89.

85   "Stockholm species": Elizabeth Kolbert, *Under a White Sky: The Nature of the Future* (New York: Crown, 2021), 84.

85   The CPRA's ten-year contract: Coastal Protection and Restoration Authority, "Queen Bess Island Restoration."